Cambridge Elements

Elements of Paleontology
edited by
Colin D. Sumrall
University of Tennessee

THE TAPHONOMY OF ECHINOIDS: SKELETAL MORPHOLOGIES, ENVIRONMENTAL FACTORS AND PRESERVATION PATHWAYS

James H. Nebelsick
University of Tübingen

Andrea Mancosu
University of Cagliari

CAMBRIDGE
UNIVERSITY PRESS

University Printing House, Cambridge CB2 8BS, United Kingdom

One Liberty Plaza, 20th Floor, New York, NY 10006, USA

477 Williamstown Road, Port Melbourne, VIC 3207, Australia

314–321, 3rd Floor, Plot 3, Splendor Forum, Jasola District Centre, New Delhi – 110025, India

103 Penang Road, #05–06/07, Visioncrest Commercial, Singapore 238467

Cambridge University Press is part of the University of Cambridge.

It furthers the University's mission by disseminating knowledge in the pursuit of education, learning, and research at the highest international levels of excellence.

www.cambridge.org
Information on this title: www.cambridge.org/9781108809993
DOI: 10.1017/9781108893411

First published 2021

A catalogue record for this publication is available from the British Library.

ISBN 978-1-108-80999-3 Paperback
ISSN 2517-780X (online)
ISSN 2517-7796 (print)

The Taphonomy of Echinoids: Skeletal Morphologies, Environmental Factors and Preservation Pathways

Elements of Paleontology

DOI: 10.1017/9781108893411
First published online: August 2021

James H. Nebelsick
University of Tübingen

Andrea Mancosu
University of Cagliari

Author for correspondence: James H. Nebelsick, nebelsick@uni-tuebingen.de

Abstract: The study of echinoid evolution, diversity, and ecology has been influenced by the fact that they are represented by taxa showing widely differing architectural designs of their multi-plated skeletons, inhabiting a large range of marine paleoenvironments, which result in highly varying taphonomic biases dictating their presence and recognition. This Element addresses the taphonomy of echinoids and includes: a general introduction to the morphological features of echinoids that play a role in their preservation; a review of processes which play an important role in the differential preservation of both regular and irregular echinoids including predation and transport; a summary of taphonomic pathways included in actualistic studies of extant sea urchins and then reconstructed for fossil taxa; and, finally, a case study of the variation of echinoid taphonomy across a shelf gradient using the rich Miocene echinoid fauna of Sardinia.

Keywords: Sea Urchins, Taphonomy, Preservation, Miocene, Sardinia

ISBNs: 9781108809993 (PB), 9781108893411 (OC)
ISSNs: 2517-780X (online), 2517-7796 (print)

Contents

1 General Introduction

Throughout the history of studying fossil echinoids, researchers have had to deal with preserved remnants that are incomplete. Although these fossils may be common, almost all lack many of their original elements, are curiously incomplete or crushed, or even consist of only single elements. The fact that fossil sea urchins do not represent complete individual organisms was already apparent in early descriptions and illustrations of fossilized sea urchins. Agostino Scilla (1670), in his remarkable discourse on the true nature and origin of fossil remains, includes superb illustrations of fragmented and in situ crushed echinoid specimens (see also Romano 2013, Findlen 2018). The Welsh naturalist Edmund Lhuyd, known as Luidius (1699), as remarked in Baier (1708) and Bantz (1969), noted that of the more than 400 specimens of fossil urchins in his collection, not a single one still retained its spines. Comprehensive monographs on important fauna are based solely on disarticulated material, for example, Bather's monograph of Hungarian Triassic echinoids from Bakony (Bather 1909), among others.

Taphonomy plays a key role in the representation of echinoderms in the fossil record, dictating the taxonomic rank that can be identified, allowing evolutionary lineages to be established, and studies pertaining to diversity and disparity to be made. The taphonomy of echinoids has been included in research into the broad context of mass extinctions and recovery (Smith 1990), biogeographic patterns (Carter & McKinney 1992), time averaging (Kowalewski et al. 2018), paleoecological interpretation (e.g., Ernst 1969, 1970, Smith et al. 1995, Kroh & Nebelsick 2003, Smith & Rader 2009, Thompson & Ausich 2016, Thompson et al. 2015, see section 6 below), and quantification of taphonomic variables (Grun & Nebelsick 2016). Echinoids are important for the interpretation of *Lagerstätten* (Seilacher 1970, Seilacher et al. 1985, Brett & Seilacher 1991, Brett et al. 1997) in which the spectacular preservation of echinoids, along with other biota, indicate special environmental sedimentary conditions. These deposits include tempestites and obrution deposits in both shallow and deeper water (Kutscher 1970, Rosenkranz 1971, Hess 1972, Radwański & Wysocka 2001, Wysocka et al. 2001, Radwański & Wysocka 2004, Schneider et al. 2005, Thuy et al. 2011) as well as lithographic limestones (Bantz 1969, Roman & Fabre 1986, Roman et al. 1991, 1994, Roman 1993, Bourseau et al. 1994, Grawe-Baumeister et al. 2000, Chellouche et al. 2012, Peyer et al. 2014). Additionally, taphonomic pathways influence the extent to which echinoids contribute as sedimentary particles to the rock record, especially within mass accumulations (e.g., Nebelsick & Kroh 2002, Belaústegui et al. 2012, Mancosu & Nebelsick 2013, 2015, 2017a). Insights into the taphonomy of echinoderms are often embedded, if not hidden, within other biological and geological studies, though there have been some reviews regarding this (see Lewis 1980, Donovan 1991, 2003, Brett et al. 1997, Ausich 2001, Nebelsick 2004).

Studying the taphonomy of echinoids is thus of great importance with respect to interpreting their presence for a wide range of applications. Understanding the preservation potential of echinoid remains is paramount for their inclusion within studies with respect to evolution, biodiversity, paleoecology, sedimentation events, and other subjects.

2 Multi-plated Skeletons and Regeneration

There are two important caveats concerning the taphonomy of echinoderms that distinguish them from most other benthic marine organisms: 1) the presence of hierarchically organized, multi-plated skeletons consisting of high-magnesium calcite stereom; and 2) the loss of attached appendages, through autotomy and sub-lethal predation leading to the continuous production of skeletal elements during the life span of a single organism.

Echinoids possess multi-plated, hierarchically organized skeletons (e.g., Smith 1984, Nebelsick et al. 2015) containing numerous elements of different shapes and sizes ranging from miniscule spicules to massive spines. These elements include: 1) ambulacral and interambulacral plates; 2) ocular and genital plates including the madreporite; 3) plates within the peristomal and periproctal membranes; 4) elements of the jaw apparatus including hemipyramids, rotulae, epiphyses, compasses, and teeth; 5) spines in a great many variations and sizes; 6) supporting elements in the pedicellaria within valves and rods in their stalks; 7) rosettes of adhesive tube feet; and, finally, 8) various calcareous spicules within the body wall. In exceptional cases, articulated echinoids complete with spines, jaw elements, and pedicellaria can be preserved, and separated elements can also be recognized in the fossil record though to different taxonomic levels. The presence of plates and spines can be used to distinguish echinoid presence and distribution (e.g., Donovan 1991, 2003, Gordon & Donovan 1992, Nebelsick 1992a, 1992b, Dynowski 2012, Thompson & Denayer 2017). Other remnants recovered from the fossil record include isolated pedicellaria valves, typically within the context of micropaleontological studies (e.g., Mortensen 1934, 1937, Geis 1936, Blake 1968, Krainer et al. 1994, Mostler 2009), and even partially preserved tube feet rosettes (Mortensen 1937).

Similarly to ecdysis by crustaceans and leaf abscission by plants, a single echinoid will produce more hard parts than its constituent skeletal elements at a single point in time. Loss of skeletal elements during life can be caused by

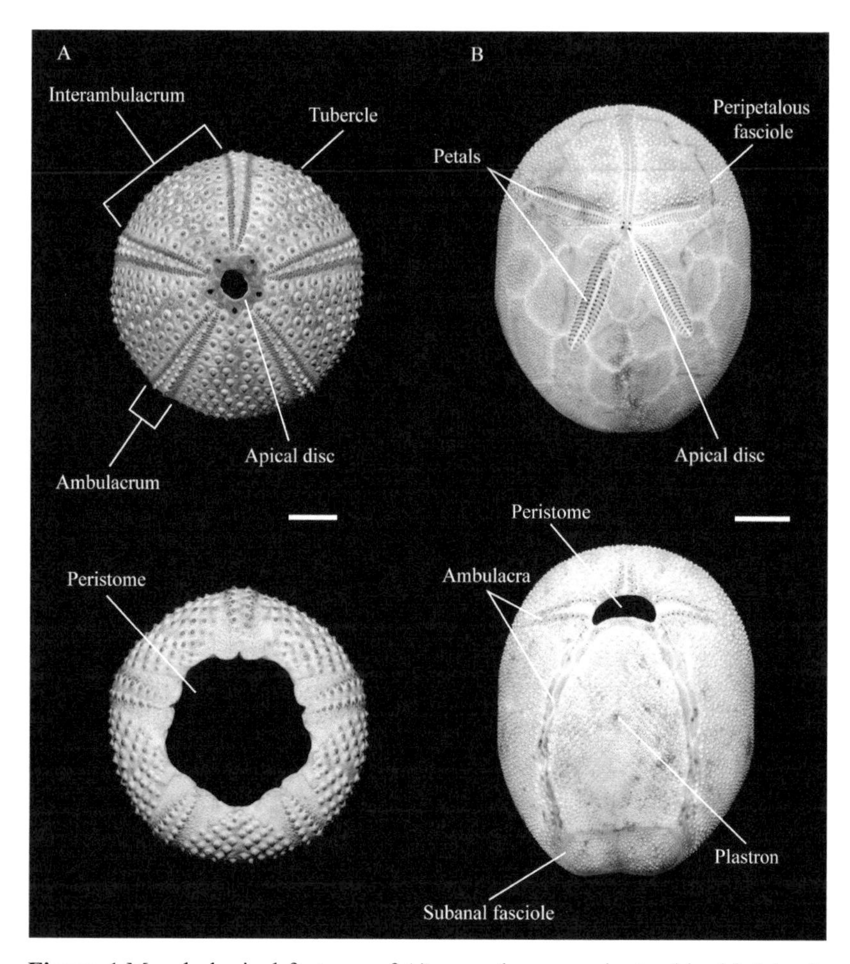

Figure 1 Morphological features of A) a regular camardont echinoid *Arbaxia lixula* and B) an irregular spatangoid echinoid *Brissus unicolor*, both from the Mediterranean Sea. The top row shows aboral (top) views of the test; the lower row shows oral (lower) views of the test. Scale bars = 1 cm.

mechanical breakage, disease, parasitism, sub-lethal predation (Lawrence & Vasquez 1996), and autotomy (Prouho 1887, Märkel & Röser 1983). Many elements such as spines, pedicellaria, and tube feet can be regenerated (Cutress 1965, Ebert 1967, 1988, Heatfield 1971, Dubois & Ameye 2001) and, thus, produce further calcitic elements. Sub-lethal predation on the tests of echinoids, especially clypeasteroids, can heal, leaving characteristic wounds in the test that can be recognized in the fossil record (see section 4). Furthermore, echinoid teeth are continuously renewed in order to compensate for abrasion at the biting surface (see Ellers & Telford 1996).

In summary, there are two characteristic features of echinoderms that have a large influence on their presence in the fossil record. The first is the presence of a multi-plated skeleton that can disarticulate to various degrees during taphonomic processes. The second is the fact that regeneration can lead to increased production of skeletal elements during the life span of an individual organism.

3 Influence of Test Architecture and Environment

Taphonomic processes determine preservation patterns of once living animals and their final representation in the rock record. There are numerous factors leading to differential preservation of echinoids, some characteristic of echinoderms as a whole, others limited to sea urchins generally, or even to specific taxa.

The general distribution of regular echinoids on hard substrates in high-energy environments and the presence of irregular echinoids within soft sediments leads to a better preservation potential for irregular echinoids (Ernst & Seibertz 1977, Kier 1977). This depends to a large extent on what type of remains are being considered: complete fossils with spines attached, preserved tests, or single fragments (see Nebelsick 1996). This, in turn, not only depends on the ambient environment, but also on the durability of both soft and hard parts. Soft parts, including connective ligaments, muscle tissues, and epithelium will decay (see Smith 1984, Allison 1990, Kidwell & Baumiller 1990). In high-energy environments, echinoid remains can be restricted to fragmented portions of the test, often showing interplate breakage and abraded fragment boundaries. In addition, spines are mostly broken, with few complete examples (Mancosu & Nebelsick 2020) preserved.

Skeletal integrity depends on the presence or lack of stereomic strengthening between plates, as discussed by Smith (1990, 2005) within an evolutionary context. Test architectures range from loose plate connections in Paleozoic echinoids, echinothuroids, and diadematoids to skeletal projections between plates in more derived regular echinoids, to robust connections including internal supports, which are found in clypeasteroids. The architecture of the echinoid skeleton has been the subject of detailed studies with respect to growth trajectories, plate proliferation, biomimetic, and taphonomic applications (e.g., Seilacher 1979, Telford 1985a, 1985b, Philippi & Nachtigall 1996, Zachos 2009, Grun et al. 2016, 2018, Grun & Nebelsick 2018a, 2018b, 2018c, Perricone et al. 2021).

There are thus numerous factors leading to the differential preservation potential of echinoids through time. These include intrinsic factors of skeletal architectures and plate connections as well as extrinsic environmental factors such as water energy and substrate conditions.

4 Predation and Parasitism as Taphonomic Agents

Both predation and parasitism can alter the skeletons of echinoids and, thus, represent important taphonomic processes. Their identification in the fossil record can be used to record predator/prey relationships through time (see Kowalewski & Nebelsick 2003, Petsios et al. 2021). Both regular and irregular echinoids serve as prey to a variety of predators, including numerous marine invertebrates, but also vertebrates including, among others, birds, sea otters, and humans. There is a large body of biological literature pertaining to echinoid predation (see chapters on specific taxa in Lawrence 2020) because echinoid predation plays a key role in structuring communities (e.g., Hendler 1977, Estes et al. 1978, McClanahan 1988, 1995, 1998, Sala & Zabala 1996, Guidetti & Mori 2005, Young & Bellwood 2011, Johansson et al. 2013). Predation on echinoids can lead to a range of different preservation styles (Nebelsick 1999b), some of them potentially recognizable in the fossil record including: 1) no signs at all (for example, predation by starfish or predators that enter the peristomal or periproctal membrane) (see also Kidwell & Baumiller 1990); 2) non-distinct wounds on the test that cannot be assigned to any specific predator, as for predation by many fish, decapod, and bird species; 3) distinct characteristic wounds that can be attributed to specific predators including gastropods and some fish; and 4) the total destruction of the test into non-distinct fragmentary skeletal elements (most durophagous predators).

4.1 Holes and Pits

There are a host of parasites that also affect echinoids (Jangoux 1984), some of them leaving distinct traces on the skeleton. Various traces attributed to parasites include galls in tests and spines (see the compilations in Donovan 2015, Belaústegui et al. 2017). Malformations include pinching of ambulacral pore rows present in fossils (e.g., Tavani 1935, Roman 1952, 1953, Marcopoulos-Diacantoni 1970, Abdelhamid 1999, Zamora et al. 2008) as well as pits and holes in the tests (Grun et al. 2020). Pinnotherid crabs infest echinoids, in sand dollars shaving off patches of the minute spines (see Martinelli Filho et al. 2014, Nebelsick 2020). Parasitic gastropods enter through ambulacral pores or produce holes in the test. Detailed description and ichnotaxa are especially well represented for Cretaceous echinoid tests (Jagt et al. 2007, Donovan et al. 2008, 2014, 2017, Donovan & Jagt 2013, Neumann & Wisshak 2006, 2009, Neumann et al. 2008) and include a wealth of holes and pits caused by, among others, foraminifera, gastropods, and brachiopods. For a summary of the protracted discussions with respect to the naming and interpretation of these ichnotaxa, see Belaústegui et al. (2017).

Cassid gastropods are also known to drill characteristic holes in the tests of echinoids. These possess irregular outlines due to the presence of skeletal features such as tubercles and ambulacral holes (see the compilation in Nebelsick 1999b, Nebelsick & Kowalewski 1999, Grun et al. 2014, Grun 2017). The presence of holes in the small clypeasteroids *Echinocyamus* and *Fibularia* has been studied in detail because these taxa can occur in large numbers, allowing for the quantification of predation intensity and site selectivity (Nebelsick & Kowalewski 1999, Ceranka & Złotnik 2003, Złotnik & Ceranka 2005, Grun et al. 2014, 2017, Meadows et al. 2015, Grun 2017). There has been much discussion as to whether predatory and parasitic gastropod attacks can be distinguished (see discussion in Kier 1981, Gibson & Watson 1989, Rose & Cross 1993, Cross & Rose 1994, Ceranka & Złotnik 2003, Donovan & Pickerill 2004, Meadows et al. 2015, Grun et al. 2017). In an actuopaleontological study on *Meoma ventricosa* from San Salvador Island, Bahamas, Tyler et al. (2018) demonstrated that gastropod predation can in fact be conducive to echinoid preservation. As cassid predation represents a relatively non-destructive process, it thereby effectively removes the individual echinoid from more destructive predation events.

4.2 Durophagous Predation

Although durophagous predation has obvious implications for sea urchin preservation (see Nebelsick 1999b, Kowalewski & Nebelsick 2003, Zatoń et al. 2007, Borszcz & Zatoń 2013), there are few studies specifically addressing its taphonomic implications. Despite the fact that decapods are known to prey on echinoids, their prey handling and resultant traces have not been studied in detail. The commonly occurring, highly irregular outline of sand dollars (Fig. 2B-D) has been attributed to non-lethal crab predation (Merrill & Hobson 1970, Birkeland & Chia 1971, Lawrence & Vasquez 1996). The wounds are healed without restoration of the original test outline allowing these interactions to be recognized in the fossil record (e.g., Tasnádi-Kubaska 1962, Marcopoulos-Diacantoni 1970, Zinsmeister 1980, Ali 1982, Nebelsick 1999a).

Although fish predation generally destroys echinoid tests, the robust tests of clypeasteroids are stable enough to withstand total destruction and they are left with characteristic wounds (Fig. 2B) consisting of scrape marks and an eviscerated central test area (Kurz 1995, Nebelsick 1999a, 1999b). Detailed studies carried out by Grun (2016) on stingray predation on the large spatangoid *Meoma ventricosa* and the sand dollar *Leodia sexiesperforata*, and by Sievers & Nebelsick (2018) on sparid fish predation on the regular Mediterranean

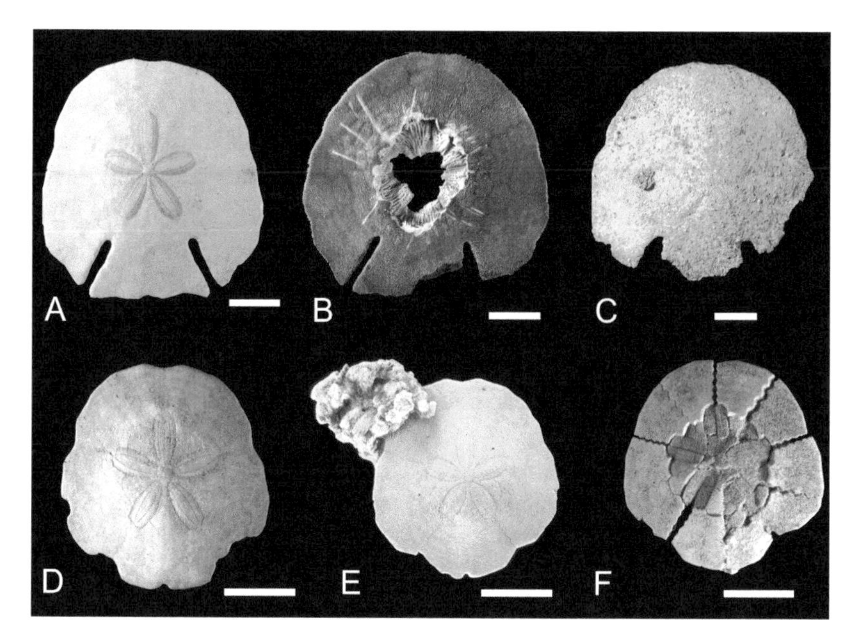

Figure 2 Comparison of taphonomic pathways of clypeasteroid sand dollars from the Recent (Northern Bay of Safaga, Egypt-top row) and Lower Miocene (Lower Austria-bottom row). A–C) *Sculpsitechinus auritus*. A) Complete denuded specimen. B) Recently fish-predated specimen still carrying spines and showing tooth marks and eviscerated central test area; note posterior sub-lethal predation. C) Heavily sub-lethally bitten test showing abrasion and encrustation. D–E) *Parmulechinus hoebarthi*. D) Denuded specimen showing non-lethal predation along the ambitus of the test. E) Post-mortem encrustation by coralline algae. F) Post-depositional compaction and implosion of the central test area with spreading along the meridional and adradial sutures. Scale bars = 2 cm.

echinoid *Sphaerechinus granularis*, outline the range of destruction left by these attacks and the possibilities of their recognition in the fossil record. A number of bird species including eider ducks, gulls (Merrill & Hobson 1970), and crows predate echinoids in shallow subtidal or intertidal sediments (see literature in Sievers et al. 2014). Predation by carrion crows, *Corvus corone*, on the spatangoid *Echinocardium cordatum* from the sandy beaches of Brittany in France produces distinct puncture marks and intraplate fragmentation allowing the birds to eviscerate the tests (Sievers et al. 2014).

In summary, both predation and parasitism represent complex ecological interactions that can profoundly influence the preservation of echinoid skeletons. The recognition of these processes depends on the presence of characteristic wound morphologies, which can vary highly among specific interactions.

5 Preservation Pathways in the Present and Past

Research on the taphonomic pathways of echinoids range from: 1) tracking the successive destruction of the test, from living organism to discrete elements and non-identifiable fragments; 2) analyzing the complex fate of echinoids serving as substrates for epi- and endobionts, which can play an important role in both the structuring of ecosystems and production of sediment; to 3) exploring the expected and unexpected excellent preservation of sea urchins within specific environments. These pathways have been explored for both extant and fossil echinoids, typically within the same study and range, from rather straightforward sequences to complex, protracted chains of processes and eventual outcomes.

5.1 Disarticulation and Fragmentation

There is a hierarchy of sequential preservation states as demonstrated in recent studies and the fossil record. These have been compiled for regular and irregular echinoids (e.g., Schäfer 1962, 1972, Ernst et al. 1973, Greenstein 1989, 1991, Allison 1990, Kidwell & Baumiller 1990; Nebelsick & Kampfer 1994). In general, there are stages of disarticulation and fragmentation: 1) complete skeletons including tests, jaws, and various appendages; 2) complete tests consisting of conjoined ambulacral and interambulacralia rows, commonly missing the apical system; 3) test fragments consisting of conjoined plates – either interplate broken along the plate boundaries or intraplate fragmentation within plates or a combination of both; 4) isolated disarticulated plates and other complete elements (spines, apical plates, jaw elements, pedicellaria valves, and rods); and 5) fragmented elements. The exact chronological sequence and intensities of these stages depends on numerous factors and differs widely between and among regular and irregular echinoids. Furthermore, some taxonomic groups have been studied in detail, while others have hardly been analyzed, especially with respect to plate thicknesses, interplate connections, soft part morphologies, and so forth.

A comparison of the taphonomic pathways of regular versus irregular echinoids was described by Schäfer (1962, 1972) from two different environments of the German Bight in the North Sea. In soft substrates, the irregular spatangoid echinoids *Echinocardium* burrow deeply in coarse sands. During high-energy events, these echinoids are either smothered by a thick layer of sediment or scoured out of the sediment, and they can be transported shoreward. The latter can form an ephemeral mass accumulation up to a meter wide and a kilometer long, which is destroyed within days or weeks. Complete specimens are thus only preserved if they are buried and not extracted because extracted dead tests are readily fragmented. The tests of regular echinoids *Echinus esculentus*, *E. acutus*, and *Strongylocentrotus droebachiensis* inhabiting hard and rocky substrates can either be destroyed by durophagous predators, such as the Atlantic wolffish, *Anarhichas lupus*, or by moving boulders, both of which lead to fragmentation. *Psammechinus miliaris* living on sandy substrates have a better chance of preservation, as they can dig themselves free from 5 cm to 20 cm depths if covered by sediment. Echinoids suffering non-destructive death have drooping spines which eventually create an apron of spines surrounding the test. After seven days, *Echinus* spines start to drop off; after 12 days the apical system disarticulates and the jaws disarticulate. The plates of regular echinoids fall apart along their sutures leading to a shell hash.

In a study of Eocene echinoids from Istria, Mitrović-Petrović (1982) suggested that transport not only accounted for abrasion and fragmentation of tests but also for mixing of fauna from different environments, a coarse-grained substrate in a higher-energy environment with *Conoclypeus* and *Echinolampas*, and finer substrate in deeper water with *Cyclaster*, *Linthia*, and *Macropneustes*. Transport and subsequent mixing of both regular and irregular echinoid remains at different scales were also reported by Nebelsick (1992a, 1996) for reefal and associated facies from shallow-water environments in the Red Sea.

In laboratory experiments on regular echinoids, Kidwell & Baumiller (1990) documented the systematic disintegration of echinoid tests. Depending on the stage of disassociation, tumbling led to the production of tests that were intact, though fractured, then large pieces, and, finally, small fragments. A threshold time dictating the rate at which echinoids disassociate is marked by the decomposition of collagenous tissues (see also Simon et al. 1990, Smith et al. 1990, Ellers et al. 1998, Johnson et al. 2002), which cross plate sutures. Because temperatures affected organic decomposition far more than anoxia, there is a potential for latitudinal as well as bathymetric gradients in the preservation of fossil echinoid faunas (Kidwell & Baumiller 1990). Undisturbed coronas can remain intact for months, which allows sufficient time for epibiont occupation (Kidwell & Baumiller 1990). Using recent echinoids as a taphonomic proxy for extinct organisms, Allison (1990) also studied the decay rates under different environmental conditions.

A number of actualistic studies on echinoids in the Caribbean by Greenstein (1989, 1990, 1991, 1992, 1993a, 1993b, 1995) compared regular echinoid taphonomy to their skeletal architecture and, furthermore, explored taphonomic overprints with respect to the fossil record. The fact that echinoids are quickly acted on by biostratinomic processes requires extraordinary circumstances to lead to their preservation. Differences in skeletal durability between different taxa were also measured experimentally. *Diadema* is particularly vulnerable to taphonomic bias with its poor preservation potential leading to a lack of an echinoderm spike in the sediment following their mass death due to disease. In general, regular echinoid distribution was not reflected by an accumulation of carcasses, but the reverse is true for irregular taxa which have a higher likelihood of preservation.

The utility of using echinoid fragments in fossil environments was shown by Gordon and Donovan (1992) to delineate facies types in the Pleistocene of Jamaica. Using disarticulated remains, Donovan (2001, 2003) demonstrated that regular echinoids were at least as diverse as irregular echinoids despite the lack of articulated tests (see also Dixon & Donovan 1998). Donovan & Gordon (1993) demonstrated how taphonomy affects the fossil record by comparing the sturdy *Echinometra viridis*, which is represented by tests along with plates and spines, to diadematoid plates that are locally very common, as detected by microscopic analysis. In a study on Lower Carboniferous echinoids, Thompson and Denayer (2017) demonstrated how using disarticulated ossicles can substantially increase our knowledge concerning the abundance and diversity of Paleozoic echinoids.

Using material from the Northern Bay of Safaga, Egypt, Nebelsick (1992a, 1992b, 1995a, 1995b, 1995c, 1996) used echinoid test finds as well as fragments from bulk samples to discern their distribution and correlate them to life habits, sediment distribution, and taphonomy. The analysis of bulk samples overcame the problem of patchy distributions as they reflect time-averaged accumulation. As demonstrated before, the preservation potentials of both regular and irregular echinoid taxa are highly differential depending on skeletal architecture and environmental factors. Regular echinoids have a much lower chance of preservation as a complete test than irregular echinoids, though their fragments can be used to detect their presence. Different preservation states of *Clypeaster humilis* (Nebelsick 2008) ranged from well-preserved specimens with spines still attached (victims of fish predation) to specimens lacking spines to those missing the apical system. Successive abrasion of the test surface, encrustations and bioerosion, and corrosion eventually lead to highly deteriorated tests. Nebelsick and Kampfer (1994) followed the disarticulation of clypeasteroids within the sediment and showed spine loss to occur within a few days. Once the structural integrity of the tests was disturbed (for example by scavenging gastropods), plate disarticulation rapidly occurred. The distribution of taphonomic processes on *Clypeaster* fragments (Nebelsick 1999c) led to the designation of taphofacies, among others, by abrasion (indicative of agitated shallow water) and encrustation (long sediment residence times in deeper water).

5.2 Encrustation and Bioerosion

Encrustation of the echinoid test is precluded by the fact that the test is usually covered by an epithelium because the skeleton is of mesodermal origin and is protected by pedicellaria (see Coppard et al. 2012). Cidaroid spines, in contrast, lose their epithelium after growth and thus are "naked," exposing their often highly sculptured spine surfaces to a host of epibionts (e.g., Hopkins et al. 2004, Linse et al. 2008, David et al. 2009). Encrusted spines from the Carboniferous (Schneider 2003, 2010) and Jurassic (Wilson et al. 2015) have been recorded. A further example of encrustation on living echinoids is by barnacles on sand dollars. Although best known for the barnacle *Balanus pacificus* encrusting the Pacific sand dollar *Dendraster excentricus*, this interaction was also recorded on other taxa (see Nebelsick 2020) and on fossil sand dollars (e.g., Philippe 1983).

Dead echinoid tests can be considered substrates of moderate size (see McKinney and Jackson 1989) and are attractive to epizoobionts (*sensu* Taylor & Wilson 2002) consisting of both encrusting and bioeroding organisms. Echinoid tests, with their variegated surface morphologies, represent relatively large and stable secondary substrates and are commonly encrusted (see Nebelsick et al. 1997, Borszcz 2012, Borszcz et al. 2013, Donovan & Jagt 2018). Infestation can have important implications for preservation by both stabilizing (by encrusting) and weakening (by bioeroding) the test. In an actualistic study following taphonomic pathways from living echinoids and dead tests from the Adriatic Sea, Ernst et al. (1973) inferred that encrustation increased the stability of the echinoids by crossing plate boundaries. This was also suggested by Kidwell and Baumiller (1990) with encrusters such as coralline algae reinforcing the skeleton and, thus, contributing to a relatively long sea-surface residence time. Encrustation of *Meoma* by oysters and barnacles was cited as a factor ensuring that these large, thin shells were capable of surviving on the sea floor (Donovan & Clements 2002).

5.3 Epizoobionts and Taphonomic Pathways

The taphonomic pathways of Cretaceous *Micraster* studied by Kudrewicz (1992) involve echinoids that are not only encrusted but also filled with fine sediment in an otherwise coarser sediment matrix. A complex sequence of events is invoked: 1) living animals; 2) death by senescence, since predation traces are missing, movement to the surface, and, thus, exposure at the sediment-water interface; 3) decay of soft tissues; 4) growth of epibionts (bivalves and polychaetes) on the exposed test surface and interior; 5) infilling of the test by fine sediments; 6) winnowing of the fine sediment followed by overturning and redeposition; 7) final burial of the specimens occurs.

In the Northern Adriatic Sea, denuded tests of the thin-shelled spatangoid *Ova canaliferus* are surprisingly stable and serve as secondary substrates after being forced to the surface during low oxygenation events and perishing (Nebelsick et al. 1997). Within two years, the tests are heavily encrusted (and thus stabilized) by a range of organisms that may or may not have fossilization potential. Together with a host of vagile organisms, multi-species clumps (Fedra et al. 1976) are formed leading to an increase in local biodiversity. These observations were used to interpret encrusted *Echinolampas* shells from the Miocene of Austria (Nebelsick et al. 1997). Encrustation by various bryozoan growth forms, barnacles, serpulid polychaetes, as well as coralline algae have differences among the cryptic oral and exposed aboral surfaces. By comparison to the recent examples, the encrusting epifauna was probably more diverse, as soft-bodied organisms and those characterized by non-calcified attachments typically cannot be directly traced on the echinoid surface. Similar encrustation of bryozoans on Middle Miocene *Echinolampas* was described by Mitrović-Petrović & Urošević-Dačić (1962) and encrustation by coralline algae (Fig. 2 F) on Lower Miocene *Parmulechinus* was described by Nebelsick (1999a).

The analysis of sclerobiont infestation of Late Cretaceous *Micraster* of northern Spain by Zamora et al. (2008) allowed complex taphonomic pathways to be distinguished. Sclerobionts include both encrustation and bioerosion. Encrusters include bivalves, polychaetes, encrusting foraminifera, and bryozoans. Bioerosion is represented by a wide range of ichnotaxa that formed diverse holes and pits. Further features include muddy infilling which can contain burrows as well as collapsed in situ fragmentation of tests. The high number of encrusted tests leads to the conclusion that the exhumed echinoid represented a stable substrate in a muddy sediment.

Several studies on bioinfestation have involved robust, thick-shelled, highly domed *Clypeaster* specimens from the Miocene of Mediterranean (e.g., Mitrović-Petrović 1972, Freneix & Roman 1979, Marcopoulos-Diacantoni 1984, Santos et al. 2003). Santos and Mayoral (2008) showed encrustation of *Clypeaster* tests by innumerous small barnacles allowing encrustation strategies of the spat to be analyzed with respect to the topography, surface roughness, and orientation of the tests. Dead *Clypeaster* shells also serve as relatively stable benthic islands for intensive endo-skeletozoan colonization by gastrochaenid bivalves (Belaústegui et al. 2013), with two modes of occurrence. First, clavate borings that are largely restricted to the echinoid stereom of the thick-walled test. Second, semi-endoskeletal dwellings cross through test walls and extend into the sedimentary infill of the empty test space, thereby producing carbonate crypts. Numerous cross-cutting, bioerosional structures of a single *Clypeaster* specimen were studied using μCT scans by Rahman et al. (2015) allowing for the intensity of infestation as well the identification of the trace-making bivalve to be determined.

5.4 Transport

Transport of echinoids in quieter-water settings during the Cretaceous led to accumulations either in shallow erosional depressions in the sea floor or in the empty living chamber of large ammonites lying on the sea floor (Ernst 1967). Storm events are known to either dislodge regular echinoids from their epibenthic habitats or scour out irregular echinoids from within their substrates. The delicate and light disarticulated spines of *Echinocardium cordatum* can be transported separately from the tests leading to spine accumulations entangled in seagrass remnants along the coast of the meso-tidally influenced North Sea coast (Schwarz 1930). Transport can lead to loosely distributed spines on bedding plane surfaces that can also indicate current flow. Mass accumulations of spines also occur with few associated tests or test remains (e.g., Moffat & Bottjer 1999), which suggests significant clast sorting during transport. Donovan & Embden (1996) and Donovan (2000) suggested downslope transport of robust echinoid tests leading to the presence of shallow water taxa in deeper water settings. For mass accumulations of echinoids, see discussion of Sardinian examples (see section 6.2.1).

5.5 Post-sedimentary Fate of Echinoid Skeletons

Compaction of the echinoid skeletons can lead to implosion of tests not completely filled by sediment as described by Müller (1957) for spatangoid echinoids in his classic treatment of biostratinomic processes. In situ crushing follows inherent weak points of the test including plate boundaries. Implosion of the central area of sand dollar within the supporting meshwork which is distributed around the ambitus and strengthens the skeleton is often observed in clypeasteroid skeletons such as those of the Lower Miocene *Parmulechinus* (Fig. 2 F) (Nebelsick 1999b). The tests of the large, thin-shelled spatangoid *Meoma* from the Pliocene of South Carolina may also be collapsed (Donovan & Clements 2002).

The high-magnesium calcite skeleton of echinoderms will be transformed into low-magnesium calcite unless embedded in very fine sediment that restricts the movement of pore water. Thus, excellently preserved Paleogene spatangoid echinoids from Texas were deposited in fine sediments and still retain original high-magnesium calcite mineralogy (Zachos 2008). The transformation to low-magnesium calcite is commonly accompanied by the production of syntaxial cements and is important for cementation of sediments, transforming them into indurated rocks, especially in non-tropical settings (e.g., Kroh & Nebelsick 2010). During diagenesis, syntaxial, high-magnesium cement commonly fills the pore space of the stereom, thus making it difficult to distinguish stereom types (Smith 1984).

Both actualistic and paleontological studies on echinoid taphonomy have revealed a wide range of taphonomic processes that affect preservation potentials. These include disarticulation, fragmentation, encrustation, bioerosion as well as factors related to transport. Taphonomic pathways can be complex and lead to either the destruction or the preservation of the skeleton. Finally, diagenetic features which accompany the transformation of the echinoid shells into fossils within the rock record must be taken into account.

6 A Case Study of Echinoid Preservation: The Miocene of Sardinia

6.1 Sardinian Echinoids

The Miocene echinoid fossil record of Sardinia offers an exemplary case study for the importance of the inclusion of taphonomic analysis for reconstructing paleoenvironments. This is due to several factors. Oligo-Miocene sediments in Sardinia (Fig. 3) have excellent exposure due to a complex geological history with active tectonics, a seasonal climate leading to both wet (erosion) and dry (limited vegetation) conditions, and the fact that it is an island featuring dramatic coastal exposures. There is a long history of study of these respective rock units allowing for a high-resolution stratigraphy and sedimentological framework to be included in the analyses.

Echinoids represent key faunal elements in Miocene sediments of Sardinia and in fact dominate the fossil record of various outcrops (Fig. 4). The echinoids are represented by numerous taxa of both regular and irregular sea urchins from a wide variety of facies types. These fauna have been included in a number of historical studies and monographs and are well represented in diverse museum collections (see references in the recent studies cited below). Despite the fact that the echinoids are largely in need of taxonomic revision (a work in progress), these monographs demonstrate the taxonomic spectrum of specific localities as well as a sense of the quality of preservation.

Miocene echinoids from Sardinia are the object of a number of recent studies summarized in this review (Mancosu et al. 2015, Mancosu & Nebelsick 2013, 2015, 2016, 2017a, 2017b, 2019) with the objective of collecting primary data on echinoids in the field with respect to: 1) assessing their diversity and quantifying their density of occurrence; 2) cataloging the state of preservation

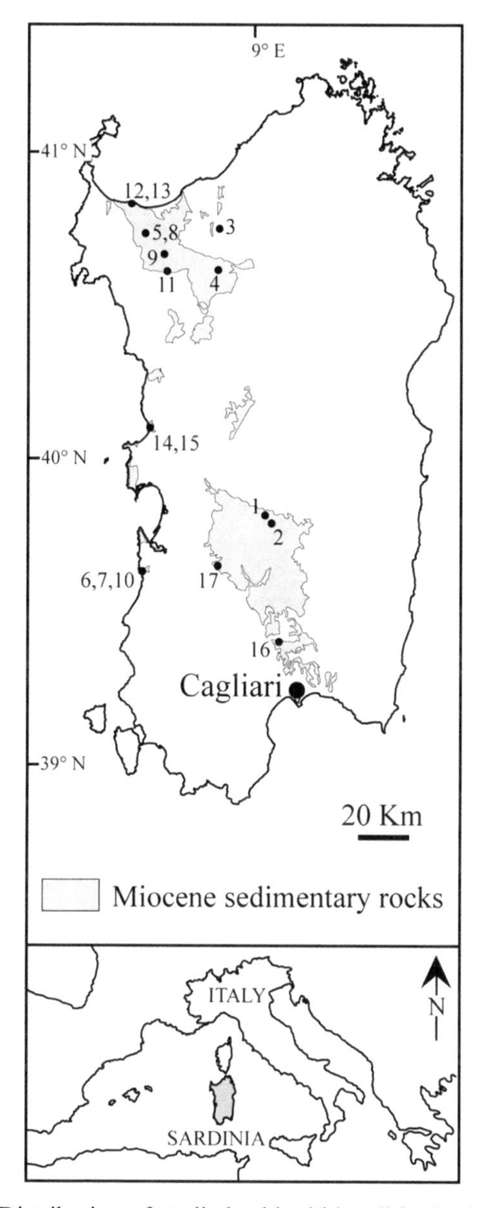

Figure 3 Distribution of studied echinoid localities in the Miocene of Sardinia (see Table 1).

including the presence of fragmentary material and spines, and quantifying specific taphonomic features including abrasion, fragmentation disarticulation, encrustation, and bioerosion; 3) incorporating other faunal and floral elements as well as sedimentological features; 4) interpreting the lifestyles of the

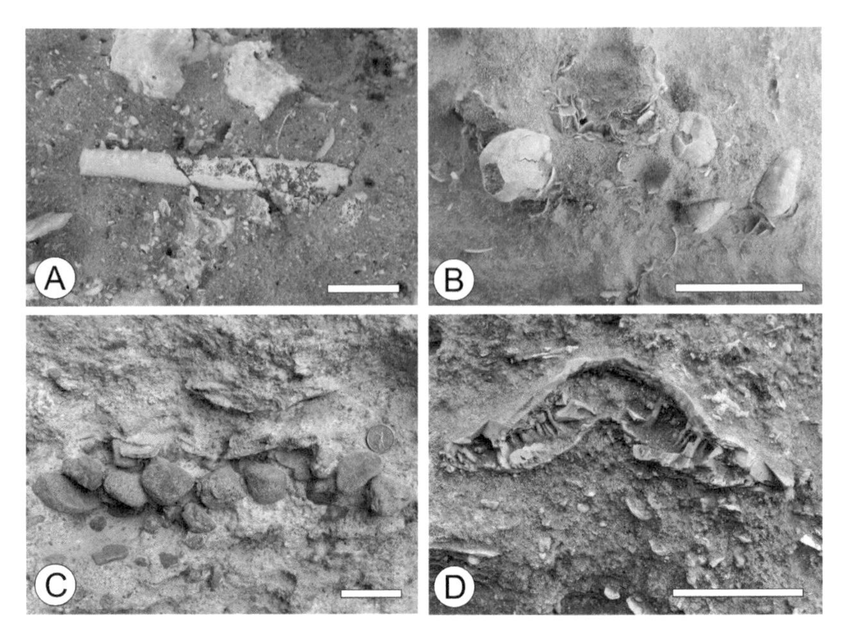

Figure 4 Representative preservation of echinoids from the Miocene of Sardinia. A) Isolated, broken spine of the cidaroid *Prionocidaris* (Porto Torres, rhodolith beds). B) Complete and fragmented thin-shelled spatangoids in part showing in situ breakage along plate boundaries (Funtanazza, fine-grained sandstones). C) Clypeasteroid echinoid tests (*Amphiope*) in cross section together with rounded cobbles (Andara). D) Eroded, highly domed, thick-shelled *Clypeaster* test showing internal support system (Bancali). Scale bars: A = 1 cm; B–D = 5 cm.

echinoids based on functional morphology as well as comparing them to equivalent or similar recent taxa with respect to skeletal architectures, species interactions, ecological distributions, and taphonomic pathways; and, finally, 5) defining and interpreting echinoid biofacies and including them in a general paleoenvironmental analysis.

6.2 Miocene Echinoid Assemblages

Altogether, 13 different localities, ranging from littoral, shallow sublittoral, sublittoral, and deep sublittoral environments from the Miocene of Sardinia are included in this analysis (Table 1). These localities are compared with respect to diversity, relative abundance of taxa, and preservation as a complete test and test fragments and/or spines. Four localities can be subdivided into different paleoenvironmental categories: Bancali, littoral to shallow sublittoral; Funtanazza, an extensive outcrop ranging from littoral to sublittoral; and Porto Torres and Santa Caterina-S'Archittu both containing fine-grained

Table 1 Table 1 Diversity, Distribution, and Preservation of Echinoids in the Miocene of Sardinia

Paleoenvironments	**L**	**L**	**L**	**L**	**L**	**L**	**SH**	**SH**	**SH**	**SH**	**SL**	**SL**	**SL**	**SL**	**SL**	**SL**	**DS**
References	**A**	**A**	**A**	**B**	**C**	**D**	**D**	**C**	**C**	**D**	**B**	**E**	**E**	**F**	**F**	**G**	**G**
Taxa / Localities	**1**	**2**	**3**	**4**	**5**	**6**	**7**	**8**	**9**	**10**	**11**	**12**	**13**	**14**	**15**	**16**	**17**
Diadematoids																	
Diadema													*F, S*		**F, S**		
Cidaroids																	
Eucidaris									*S*				**F, S**				
Prionocidaris						*F, S*	*F, S*						**F, S**				
Stylocidaris										**T, F, S**							
Tylocidaris						*F*	*S*										
Camarodonts																	
Genocidaris						**T, F**	**T, F**		*T*	**T, F**		**T, F**	**T, F**	**T, F**			
Brochopleurus												*T, F*					
Schizechinus															**T, F**		
Tripneustes							**F**					*F*			**T, F**		
Phymosomatidae																**T, F, S**	**T, F, S**
Echinoneioids																	
Koehleraster								*T*				*T, F*					
Echinolampadoids																	
Hypsoclypus						**T, F**	**T, F**			*T*	*T*	*F*					
Echinolampas	*T, F*					**T, F**	**T, F**	**T, F**	**T, F**								

Table 1 (cont.)

Clypeasteroids																	
Clypeaster	**T, F**		*T, F*	<u>**T, F**</u>	**T, F**	*T, F*	*T, F*	**T, F**	<u>**T, F**</u>		<u>**T, F**</u>	*T, F*	*T, F*	**T, F**			
Amphiope		<u>**T, F**</u>	<u>**T, F**</u>	<u>**T, F**</u>	<u>**T, F**</u>												
Parascutella	<u>**T, F**</u>	*T, F*	*T, F*														
Echinocyamus						**T, F**	<u>**T, F**</u>			**T, F**		<u>**T, F**</u>	**T, F**	<u>**T, F**</u>	**T, F**		
Spatangoids																	
Agassizia			**T, F**														
Brissopsis										<u>**T, F**</u>		<u>**T, F**</u>	**T, F**	<u>**T, F**</u>	**T, F**		<u>**T, F, S**</u>
Echinocardium															*T, F*		
Faorina										*T, F*							
Hemipatagus												**T, F**		**T, F**	*T, F*		
Holanthus												**T, F**					
Lovenia							**T, F**			**T, F**		*T, F*		**T, F**			
Mariania										*T, F*							
Metalia																	
Pericosmus												*T*					
Opissaster												**T, F**		*T*			
Ova											**T, F**	<u>**T, F**</u>	**T, F**	**T, F**	<u>**T, F**</u>		
Schizaster							**T, F**			**T, F**		**T, F**					
Spatangus						**T, F**	**T, F**			<u>**T, F**</u>							
Trachypatagus							*T*										

Localities				Paleoenvironments		Relative Abundance	
1	Duidduru	**13**	Porto Torres: RB	**L**	Littoral	*Rare*	
2	Cuccuru Tuvullau	**14**	Sa. C.-S'Arch.: FGS	**SH**	Shallow sublittoral	**Common**	
3	Chiaramonti	**15**	Sa. C.-S'Arch.: RB	**SL**	Sublittoral	**Abundant**	
4	Ardara	**16**	Ussana	**DS**	Deep sublittoral		
5	Bancali: L	**17**	Villanovaforru			**References**	
6	Funtanazza: L			**Preservation**		**A**	Mancosu & Nebelsick 2013
7	Funtanazza: SH			**T**	Complete tests	**B**	Mancosu & Nebelsick 2015
8	Bancali: SH	Sa. C-S'Arch. = Santa Caterina-S'Archittu		**F**	Test fragments	**C**	Mancosu & Nebelsick 2017a
9	Usini			**S**	Spines	**D**	Mancosu & Nebelsick 2016
10	Funtanazza: SL	FGS: Fine-grained sandstones				**E**	Mancosu & Nebelsick 2017b
11	Ittiri	RB: Rhodolith beds				**F**	Mancosu & Nebelsick 2019
12	Porto Torres: FGS					**G**	Mancosu et al. 2015

sandstones and intercalated rhodolith beds. Thus, there are 17 different stratigraphic units that can be compared.

Table 1 reveals a number of interesting details. The littoral facies is dominated by clypeasteroids that are preserved as complete tests and fragments. Clypeasteroids are represented by the various growth forms of the genus *Clypeaster* in shallow sublittoral and sublittoral zones, again restricted to tests and fragments. The presence of cidaroids (test fragments and spines) and camarodonts in the sublittoral of Funtanazza may be indicative of sea grass beds, especially those with the small echinoid *Genocidaris*. The highest diversity is in the shallow sublittoral and sublittoral, especially in the fine-grained sandstones of Funtanazza and Porto Torres. Complete echinoids with spines are found in deeper water. There are four exceptional sedimentary environments with different types of taxonomic diversity and preservation styles.

6.2.1 Mass Accumulations of Clypeasteroids

Description: Mass occurrences of clypeasteroid echinoids in shore environments (Fig. 4) are found in Duidduru (Genoni), Cuccuru Tuvullau (Nuragus), Monte Sa Loca (Chiaramonti), and Ardara (Table 1). The clypeasteroid-bearing deposits are characterized by a low diversity and are dominated by the genera *Amphiope* and *Parascutella* with subordinate *Clypeaster* present with various morphotypes, and other accompanying taxa, such as the spatangoid *Agassizia* and the echinolampadoid *Echinolampas*. The echinoid remains, always lacking spines, can range from well-preserved tests showing surface features such as tuberculation and ambulacral pores to poorly preserved, abraded specimens. They include complete tests and variously sized test fragments, which range from those representing up to half of the skeleton, to pie-shaped portions of the test, to very small fragments. Complete tests occur in high density reaching a hundred individuals per m^2. Echinoid tests and fragments are loosely to densely packed and can be oriented from vertical to concordant to the bedding plane. Test imbrication also occurs. The specimens commonly show evidence of encrustation by barnacles, serpulids, and bryozoans along with traces of boring activities by macro-endolithic organisms and predation. Post-depositional features consist of grain indentation of the test surface and radial cracking resulting from sediment loading.

Interpretation: Mass accumulation of clypeasteroid echinoids in shallow water, higher energy environments are facilitated by two factors: 1) Clypeasteroids can occur in great densities in mass accumulations (see compilation in Nebelsick & Kroh 2002, Belaústegui et al. 2012, Mancosu & Nebelsick

2015, 2017a), 2) Clypeasteroids include the most taphonomically robust echinoids (Fig. 4D) with interlocking plates and internal supports (Kier 1977, Seilacher 1979, Telford 1985a, Grun & Nebelsick 2016, 2018a, 2018b, 2018c). These accumulations can have large differences with respect to relative abundance of taxa, details of sedimentary setting, and the origin of the deposits; but in general they have low taxonomic diversity.

6.2.2 Spatangoids in Highly Bioturbated Sediments

Description: Fine-grained sedimentary successions of sublittoral environments cropping out in Porto Torres (Fig. 4B) and Santa Caterina di Pittinuri-S'Archittu are dominated by spatangoid echinoids and the minute clypeasteroid *Echinocyamus*. Among spatangoid echinoids, the genera *Schizaster*, *Ova*, *Brissopsis*, *Metalia*, *Opissaster*, *Holanthus*, *Lovenia*, *Hemipatagus*, *Echinocardium*, and *Pericosmus* were found. The clypeasteroid *Clypeaster marginatus* also occurred. Additionally, the echinoneid *Koehleraster* and small regular trigonocidarids were also present. Echinoid remnants occur in a wide spectrum of states of preservation and consist of both complete tests lacking spines and variously sized test fragments with both inter- and intraplate fragmentation. The echinoids can have exquisitely preserved test surface details with no evidence of abrasion and encrustation. The echinoid remains are not homogeneously distributed within the deposit and range from loosely packed and dispersed to very densely packed with complete tests reaching densities of 15 individuals/m^2 on exposed rock surfaces. The orientation of complete tests ranges from concordant to perpendicular (specimens lying on their sides) to the bedding plane. Bioerosion is present as circular drillholes attributed to *Oichnus simplex* Bromley, 1981. The fine-grained deposits are intensely bioturbated by large, branched burrows, like those of *Thalassinoides*, that are commonly filled by coarse biogenic material consisting predominately of echinoid test fragments and other shell remains.

Interpretation: The spatangoid assemblages originated due to a combination of ecological and environmental/sedimentological factors. Their gregarious behavior and their habitat in quiet sublittoral environments affected by bioturbation and episodes of rapid deposition are the main factors leading to these spatangoid-bearing deposits. Although most spatangoids recognized herein lived infaunally, their mode of life does not appear to play a prominent role in the preservation of either shallow- or deep-burrowing forms. The co-occurrence of different irregular echinoids within each assemblage suggests resource partitioning among infaunal deposit feeders. Bioturbation, by thalassinid shrimp and by spatangoids, is an important factor affecting the preservation potential of infaunal spatangoid echinoids because it is a source of shell breakage.

6.2.3 Regular Echinoids with Spines

Description: Diadematoids and cidaroids were preserved within rhodolith beds, which are intercalated to the fine-grained sedimentary successions as discrete layers in Santa Caterina and Porto Torres. Remains of the diadematoid *Diadema* occur typically as isolated ambulacral and interambulacral plates, although test fragments comprised of several articulated plates and associated spines can be present as long segments and fragments. Isolated Aristotle's lantern elements ascribed to these diadematids consist of large demipyramids, rotulae, and grooved teeth.

The cidaroids *Prionocidaris* (Fig. 4A) and *Eucidaris* are represented by test fragments that consist of adjoining ambulacral and interambulacral plates and isolated plates. Spine remains of both genera are also abundant and, in the case of *Prionocidaris*, are commonly encrusted by bryozoans. The regular toxopenustid echinoids *Tripneustes* and *Schizechinus* are also present in rhodolith beds. Cidaroids also occur commonly in Funtanazza both in coarse-grained facies of shallow water origin, where fragmented test and spine remains of *Prionocidaris* and *Tylocidaris* occasionally occur, and fine- to medium-grained, well-sorted sandstones in sublittoral environments where the cidaroid *Stylocidaris* was found in great abundance. A single articulated cidaroid find (*Stylocidaris*) probably represents burial in a storm event. The spatangoids *Schizaster*, *Spatangus*, *Brissopsis*, and *Lovenia* also occur.

The echinoid remains are randomly dispersed and do not form dense aggregations. Cidaroid echinoids have different states of articulation ranging from denuded coronas lacking the apical disc and spines to complete double interambulacral rows with adjoining single plate ambulacral rows and isolated interambulacral plates. Disarticulated and broken spines are also very common. The cidaroid echinoid tests are exquisitely preserved and have fine surface details, such as scrobicular and miliary tuberculation and the ornamentation of the spines. There are no signs of encrustation, bioerosion, and predation on echinoid tests.

Interpretation: Diadematoids have a fragile test and tend to dissociate rapidly when subjected to post-mortem transportation and reworking (Smith, 1984; Greenstein, 1989, 1991, 1993a, 1993b). Preservation of diadematoids requires special environmental circumstances, such as low levels of water agitation and episodes of rapid deposition with no subsequent reworking. This is reflected by the fossil record of diadematoids in the Miocene of Sardinia, which is extremely poor and seems restricted to specific lithofacies in sublittoral environments.

Cidaroids seem slightly more resistant to taphonomic processes than diadematoids, as discussed by Kidwell and Baumiller (1990) and Greenstein (1991, 1992). This is consistent with the fact that in the Miocene deposits of Sardinia they can occur commonly as fragmented material with spines predominant both in shallow water as well as in deeper sublittoral environments. Preservation of intact tests, however, seems restricted to quiet sublittoral environments with soft substrates.

6.2.4 Complete Echinoids from Deeper Water

Description: Monospecific mass occurrences of both regular phymosomatid echinoids and the spatangoid *Brissopsis* occur in deeper shelf environments of Gennas (Villanovaforru). In both cases, echinoid remains are exquisitely preserved as external and internal molds and include different states of disarticulation. Preservation ranges from complete tests preserved together with their spine canopies to isolated ambulacral and interambulacral plates and disarticulated spines, which are usually completely preserved from the base to the tip with only a few fragmented portions of the shaft present. In regular echinoids, the Aristotle's lantern can be observed through the peristome. Larger fragments and isolated plates do not show intraplate fragmentation.

Echinoid remains are very well preserved and fine details, such as the crenulation of the tubercles, stereom surface details, and the longitudinal striations of the spines, are present. Complete echinoid tests are concentrated in single bedding planes and lie parallel to bedding either on their oral or aboral surface. The spines lie concordant to slightly oblique to bedding. The echinoid deposit ranges from matrix to shell-supported and can be described as a densely to loosely packed deposit. There is no evidence of encrustation, bioerosion, or predation. Post-depositional compaction resulted in test flattening.

Interpretation: The mass occurrences are interpreted to be the result of both the gregarious behavior of these echinoids as well as sedimentary events leading to their excellent preservation. The echinoid deposits point to a storm-dominated shelf environment. The phymosomatid assemblage represents rapid burial through obrution of a highly dense, freshly dead community. The *Brissopsis* assemblage represents a parautochthonous accumulation of a community living at high density. Although well-preserved mass accumulations of regular echinoids and thin-shelled spatangoids can occur in different depositional settings, they have higher preservation potential in deeper water locations as these are low-energy environments with rapid influx of fine-grained sediments and the absence of reworking, winnowing, and biological disturbance.

6.3 Echinoid Presence, Preservation, and Taphonomic Bias

The overarching question is to what extent taphonomic bias affects the presence and recognition of echinoids in order to better interpret their paleoecology. A general, composite interpretation of the change of physical factors, the relative abundance of echinoids, and preservation potentials along a shelf gradient ranging from the shallow littoral to outer sublittoral is shown in Figure 5. The intensity of reworking and subsequent amalgamation of echinoid material related to tempestite frequency decreases with distance from the shore to deeper water. In contrast, the

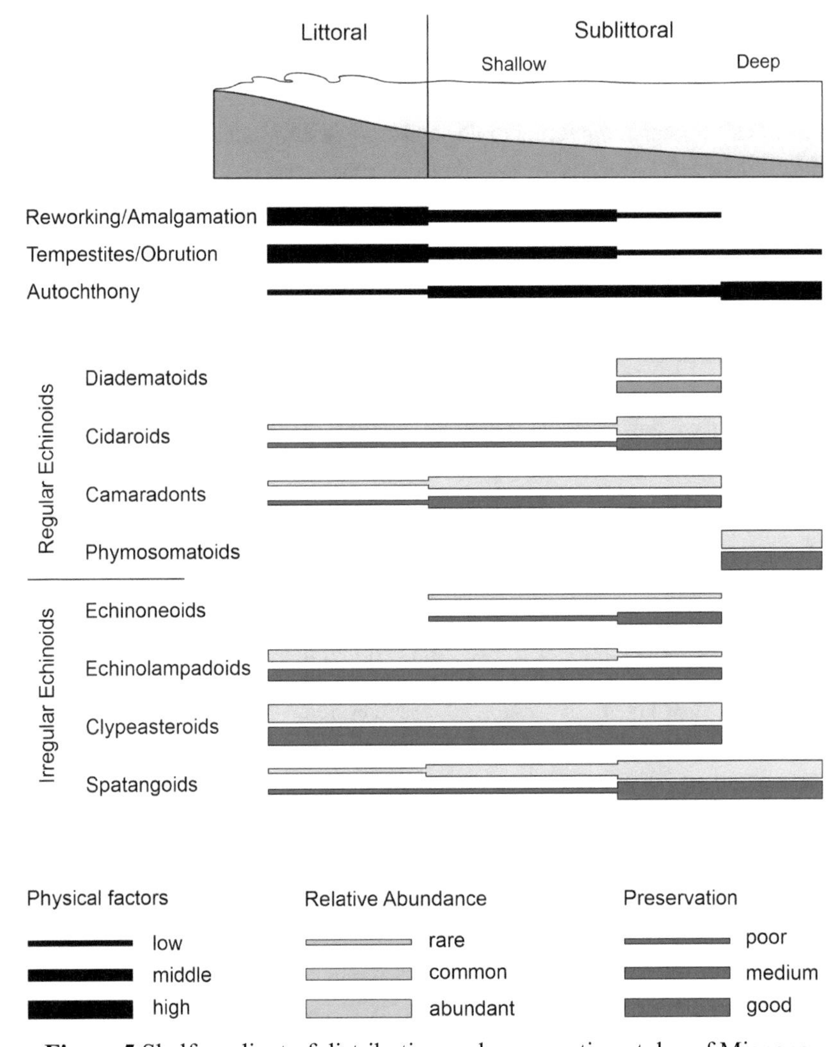

Figure 5 Shelf gradient of distribution and preservation styles of Miocene echinoids from Sardinia.

presence of autochthonous remains become more common with depth. In some cases, a lack of echinoid presence is dictated by environmental restrictions; in others, a taphonomic bias can be assumed.

Very few echinothuroids have been recognized in the Cenozoic fossil record. They are known from the Morozaki Formation in Japan (Mizuno 1993, Amemiya et al. 1994), in which they are preserved within turbiditic sequences. Diadematoids are restricted to a single facies type within relatively quiet-water fine-grained sediments. The loose plate connections and brittle spines of diadematoids pose a taphonomic bias against representation in more shallow water facies (see discussion in Mancosu and Nebelsick, 2019).

Cidaroids and camaradonts have a wide distribution with respect to depth and facies types. The massive spines of cidaroids allow for preservation in shallow water. The general lack of complete tests can be attributed to their lack of strong plate interconnections. The more robust camarodonts with complete tests along with fragments (but not spines) are in almost all of the facies. The one exception is the phymosomatid echinoid, the presence and preservation of which in deeper water is a consequence of both facies restriction and a fortuitous sedimentary event.

Irregular echinoids also have very different distributions. The common occurrence and preservation of complete clypeasteroid tests in various facies can be directly correlated to their robust test architecture. This distribution is mirrored by the less common, but also thick-shelled echinolampadoids. Echinoineoids have a more restricted presence and their taphonomic bias is difficult to interpret as there are few studies with respect to their ecology (see Rose 1976) and preservation potential. The generally better representation of spatangoids in deeper water also reflects both their distribution and fossilization potential.

7 Conclusions and Future Work

There is much work needed concerning the taphonomy of echinoids and also for echinoderms as a whole, including how the taphonomy of echinoids affects the record of diversity through time. As summarized in this review, some taxa may be under- or overrepresented in the fossil record depending on skeletal architecture, the environment in which they occur, and the specific taphonomic pathways that are taken. Although there are some exceptions (e.g., Beu et al. 1971, Banno 2008) some important taxa, such as spatangoids, have been poorly studied with respect to skeletal architecture and experimental testing of test strength. This is also true for members of the echinoineoids and echinolampadoids. Despite some early studies (Chave 1964) echinoids are not as well studied as other echinoderms as far as agitation and disassociation are concerned, echoing Gorzelak and Salamon (2013). Many predator/prey interactions are known to occur, but how they affect the preservation of echinoids and whether these kinds of attacks can potentially be recognized in the fossil record is still open to discussion. Such knowledge will be of increasing importance when trying to reconstruct the community relationships of echinoids in fossil ecosystems and how they have evolved through time.

References

Abdelhamid, M. A. F. (1999). Parasitism, abnormal growth and predation on Cretaceous echinoids from Egypt. *Revue de Paléobiologie de Genève*, **18**, 69–83.

Ali, M. S. M. (1982). Predation and repairing phenomena in certain clypeasteroid echinoid from the Miocene and Paleocene epochs of Egypt. *Journal of the Paleontological Society of India*, **27**, 7–8.

Allison, P. A. (1990). Variation in rates of decay and disarticulation of Echinodermata: implications for the application of actualistic data. *Palaios*, **5**, 432–40.

Amemiya, S., Mizuno, Y., and Ohta, S. (1994). First fossil record of the family Phormosomatidae (Echinothurioida: Echinoidea) from the Early Miocene Morozaki Group, Central Japan. *Zoological Science*, **11**, 313–17.

Ausich, W. I. (2001). Echinoderm taphonomy. In M. Jangoux and J. M. Lawrence, eds., *Echinoderm Studies 6*. Lisse: A.A. Balkema, pp. 171–227.

Baier, J. J. (1708). *Nürnbergische Fossilkunde*. Nürnberg: Wolfgang Michael, reprinted in *Erlanger Geologische Abhandlungen*, **29**, 1–133.

Banno, T. (2008). Ecological and taphonomic significance of spatangoid spines: Relationship between mode of occurrence and water temperature. *Paleontological Research*, **12**, 145–57.

Bantz, H.-U. (1969). Echinoidea aus den Plattenkalken der Altmühlalb. *Erlanger Geologischer Abhandlungen*, **78**, 1–35.

Bather, F. A. 1909. Triassic echinoderms of Bakony. *Resultate des Wissenschaflichen Erforschung des Balatonsees*, **1**, 1–286.

Belaústegui, Z., Nebelsick, J. H., de Gibert, J. M., Domènech, R., and Martinell, J. (2012). A taphonomic approach to the genetic interpretation of clypeasteroid accumulations from Tarragona (Miocene, NE Spain). *Lethaia*, **45**, 548–65.

Belaústegui, Z., de Gibert, J. M., Nebelsick, J. H., Domènech, R., and Martinell, J. (2013). Clypeasteroid tests as a benthic island for gastrochaenid bivalve colonization: Evidence from the middle Miocene of Tarragona (NE Spain). *Palaeontology*, **56**, 783–96.

Belaústegui, Z., Muñiz, F., Nebelsick, J. H., Domènech, R., and Martinell, J. (2017). Echinoderm ichnology: Bioturbation, bioerosion and related processes. *Journal of Paleontology*, **91**, 643–61.

Beu, A. G., Henderson, R. A., and Nelson, C. S. (1971). Notes on the taphonomy and paleoecology of New Zealand Tertiary Spatangoida. *New Zealand Journal of Geology and Geophysics*, **15**, 275–86.

Birkeland, C. and Chia, F.-U. (1971). Recruitment risk, growth, age and predation in two populations of sand dollars, *Dendraster excentricus* (Eschscholtz). *Journal of Experimental Marine Biology and Ecology*, **6**, 265–78.

Blake, D. B. (1968). Pedicellariae of two Silurian echinoids from western England. *Palaeontology*, **11**, 576–79.

Borszcz, T. (2012). Echinoids as substrates for encrustation – review and quantitative analysis. *Annales Societatis Geologorum Poloniae*, **82**, 139–49.

Borszcz, T. and Zatoń, M. (2013). The oldest record of predation on echinoids: Evidence from the Middle Jurassic of Poland. *Lethaia*, **46**, 141–45.

Borszcz, T., Kuklinski, P., and Zatoń, M. (2013). Encrustation patterns on late Cretaceous (Turonian) echinoids from Southern Poland. *Facies*, **59**, 299–318.

Bourseau, J.-P., Bernier, P., Barale, G., et al.. (1994). Taphonomie des échinides dugisement de Cerin (Kimméridgien Supérieur, Jura Méridional, France). Implications environnementales. *Geobios,* Mémoire spéciaux, **16**, 37–47.

Brett, C. E. and Seilacher, A. (1991). Fossil-Lagerstätten: A taphonomic consequence of event sedimentation. In G. Einsele, W. Ricken, & A. Seilacher, eds., *Cycles and Events in Stratigraphy*. New York: Springer Verlag, pp. 283–97.

Brett, C. E., Moffat, H. A., and Taylor, W. L. (1997). Echinoderm Taphonomy, Taphofacies, and Lagerstätten. In J. A. Waters & C. G. Maples, eds., *Geobiology of Echinoderms. Paleontological Society Papers*, **3**. Pittsburgh: Carnegie Museum, pp. 147–90.

Carter, B. D. and McKinney, M. L. (1992). Eocene echinoids, the Suwanee Strait, and biogeographic taphonomy. *Paleobiology*, **18**, 299–325.

Ceranka, T. and Złotnik, M. 2003. Traces of cassid snails predation upon echinoids from the Middle Miocene of Poland. *Acta Palaeontologica Polonica*, **48**, 491–96.

Chave, K. E. (1964). Skeletal durability and preservation. In J. Imbrie and D. Newell, eds., *Approaches to Paleoecology*. New York: J. Wiley & Sons, pp. 377–87.

Chellouche, P., Fürsich, F. T., and Mäuser, M. (2012). Taphonomy of neopterygian fishes from the Upper Kimmeridgian Wattendorf Plattenkalk of Southern Germany. *Palaeobiodiversity and Palaeoenvironments*, **92**, 99–117.

Coppard, S. E., Kroh, A., and Smith, A. B. (2012). The evolution of pedicellariae in echinoids: An arms race against pests and parasites. *Acta Zoologica*, **93**, 125–48.

Cross, N. F. and Rose, E. P. F. (1994). Predation of the Upper Cretaceous spatangoid echinoid *Micraster*. In B. David, A. Guille, J. P. Féral, and M. Roux, eds., *Echinoderms through Time*. Rotterdam: A. A. Balkema, pp. 607–12.

Cutress, B. M. (1965). Observations on growth in *Eucidaris tribuloides* (Lamarck), with special reference to the origin of the oral primary spines. *Bulletin of Marine Science*, **15**, 797–834.

David, B., Stock, S. R., De Carlo, F., Hétérier, V., and De Ridder, C. (2009). Microstructures of Antarctic cidaroid spines: Diversity of shapes and ecto-symbiont attachments. *Marine Biology*, **156**, 159–72.

Dixon, H. L. and Donovan, S. K. (1998). Oligocene echinoids of Jamaica. *Tertiary Research*, **18**, 95–124.

Donovan, S. K. (1991). The taphonomy of echinoderms: Calcareous multi-element skeletons in the marine environment. In S. K. Donovan, ed., *The Processes of Fossilisation*. London: Belhaven Press, pp. 241–69.

Donovan, S. K. (2000). A fore-reef echinoid fauna from the Pleistocene of Barbados. *Caribbean Journal of Science*, **36**, 314–20.

Donovan, S. K. (2001). Evolution of Caribbean echinoderms during the Cenozoic: Moving towards a complete picture using all of the fossils. *Palaeogeography, Palaeoclimatology, Palaeoecology*, **166**, 177–92.

Donovan, S. K. (2003). Completeness of a fossil record: The Pleistocene echinoids of the Antilles. *Lethaia*, **36**, 1–7.

Donovan, S. K. (2015). A prejudiced review of ancient parasites and their host echinoderms: CSI fossil record or just an excuse for speculation? In K. De Baets and T. J. Littlewood, eds., *Fossil Parasites, Advances in Parasitology*, **90**, 291–328.

Donovan, S. K. and Clements, D. (2002). Taphonomy of large echinoids; *Meoma ventricosa* (Lamarck) from the Pliocene of South Carolina. *Southeastern Geology*, **41**, 169–76.

Donovan, S. K. and Embden, B. J. (1996). Early Pleistocene echinoids of the Manchioneal Formation, Jamaica. *Journal of Paleontology*, **70**, 485–93.

Donovan, S. K. and Gordon, C, M. (1993). Echinoid taphonomy and the fossil record: Supporting evidence from the Plio-Pleistocene of the Caribbean. *Palaios*, **8**, 304–06.

Donovan, S. K. and Pickerill, R. K. (2004). Traces of cassid snails predation upon the echinoids from the Middle Miocene of Poland: Comments on Ceranka and Złotnik (2003). *Acta Palaeontologica Polonica*, **49**, 483–84.

Donovan, S. K. and Jagt, J. M. W. (2013). *Rogerella* isp. Infesting the Pore Pairs of *Hemipneustes striatoradiatus* (Leske) (Echinoidea: Upper Cretaceous, Belgium). *Bulletin of the Mizunami Fossil Museum*, **34**, 73–76.

Donovan, S. K. and Jagt, J. M. W. (2018). Big oyster, robust echinoid: An unusual association from the Maastrichtian type area (province of Limburg, southern Netherlands). *Swiss Journal of Palaeontology*, **137**, 357–61.

Donovan, S. K., Jagt, J. M. W., and Goggings, L. (2014). Bored and burrowed: An unusual echinoid steinkern from the type Maastrichtian (Upper Cretaceous, Belgium). *Ichnos*, **21**, 261–65.

Donovan, S. K., Jagt, J. M. W., and Langeveld, M. (2017). A dense infestation of round pits in the irregular echinoid *Hemipneustes striatoradiatus* (Leske) from the Maastrichtian of the Netherlands. *Ichnos*, **25**, 25–29.

Donovan, S. K., Jagt, J. M. W., and Lewis, D. N. (2008). Ichnology of Late Cretaceous echinoids from the Maastrichtian type area (The Netherlands, Belgium) – 1. A healed puncture wound in *Hemipneustes striatoradiatus* (Leske). *Bulletin of the Mizunami Fossil Museum*, **34**, 73–76.

Dubois, P. and Ameye, L. (2001). Regeneration of spines and pedicellariae in echinoderms: A Review. *Microscopy Research and Technique*, **55**, 427–37.

Dynowski, J. (2012). Echinoderm remains in shallow-water carbonates at Fernandez Bay, San Salvador Island, Bahamas. *Palaios*, **27**, 181–9.

Ebert, T. A. (1967). Growth and repair of spines in the sea urchin *Strongylocentrotus purpuratus* (Stimpson). *Biological Bulletin*, **133**, 141–49.

Ebert, T. A. (1988). Growth, regeneration, and damage repair of spines of the slate-pencil sea urchin *Heterocentrotus mammilatus* (L.) (Echinodermata: Echinoidae). *Pacific Science*, **42**, 3–4.

Ellers, O. and Telford, M. (1996). Advancement mechanics of growing teeth in sand dollars (Echinodermata, Echinoidea): A role for mutable collagenous tissue. *Biological Sciences*, **263**, 39–44.

Ellers, O., Johnson, A. S., and Moberg, P. F. (1998). Structural strengthening of urchin skeletons by collagenous sutural ligaments. *Biological Bulletin*, **195**, 136–44.

Ernst, G. (1967). Über Fossilnester in *Pachydiscus*-Gehäusen und das lagenvorkommen von Echiniden in der Oberkreide NW-Deutschlands. *Paläontologische Zeitschrift*, **41**, 221–229.

Ernst, G. (1969). Zur Ökologie und Biostratinomie des Schreibkreide-Biotops und seiner benthonischen Bewohner. *Zeitschrift der Deutschen Geologischen Gesellschaft*, **119**, 577–78.

Ernst, G. (1970). Faziesgebundenheit und Ökomorphologie bei irregulären Echiniden der nordwestdeutschen Oberkreide. *Paläontologische Zeitschrift*, **44**, 41–62.

Ernst, G. and Seibertz, E. (1977). Concepts and methods of echinoid biostratigraphy. In E. G. Kauffmanm and J. E. Hazel, eds., *Concepts and Methods of*

Biostratigraphy. Stroudsburg, PA : Dowden, Hutchinson, and Ross Inc., pp. 541–66.

Ernst, G., Hähnel, W. and Seibertz, E. (1973).Aktuopaläontologie und Merkmalsvariabilität bei mediterranen Echiniden und Rückschlüsse auf die Ökologie und Artumgrenzung fossiler Formen. *Paläontologische Zeitschrift*, **47**, 188–216.

Estes, J. A., Smith, N. S., and Palmisano, J. F. (1978). Sea otter predation and community organization in the western Aleutian Islands, Alaska. *Ecology*, **59**, 822–33.

Fedra, K., Olscher, E. M., Scherubel, C., Stachowitsch, M., and Wurzian, R. S. (1976). On the ecology of a North Adriatic benthic community: Distribution, standing crop, and composition of the macrobenthos. *Marine Biology*, **38**, 129–45.

Findlen, P. (2018). Projecting Nature: Agostino Scilla's Seventeenth-Century Fossil Drawings. *Endeavour*, **42**, 99–132.

Freneix, S. and Roman, J. (1979). Gastrochaenidae endobiotes d'échinides cénozoïques (*Clypeaster* et autres). Nouvelle classification de ces bivalves. *Bulletin Muséum National d'Histoire Naturelle, Paris*, **série 4, 1**, sect. C, 4, 287–313.

Geis, H. L. (1936). Recent and fossil pedicellariae. *Journal of Paleontology*, **10**, 427–48.

Gibson, M. A. and Watson, J. B. (1989). Predatory and non-predatory borings in echinoids from the upper Ocala Formation (Eocene), north-central Florida, USA. *Palaeogeography, Palaeoclimatology, Palaeoecology*, **71**, 309–21.

Gordon, C. M. and Donovan, S. K. (1992). Disarticulated echinoid ossicles in paleoecology and taphonomy: The last interglacial Falmouth formation of Jamaica. *Palaios*, **7**, 157–66.

Gorzelak, P. and Salamon, M. A. (2013). Experimental tumbling of echinoderms – Taphonomic patterns and implication. *Palaeogeography, Palaeoclimatology, Palaeoecology*, **386**, 569–74.

Grawe-Baumeister, J., Schweigert, G., and Dietl, G. (2000). Echinoids from the Nusplinger Lithographic Limestone (Late Kimmeridgian, SE Germany). *Stuttgart Beiträge zur Natürkunde B*, **286**, 1–39.

Greenstein, B. J. (1989). Mass mortality of the West-Indian echinoid *Diadema antillarum* (Echinodermata: Echinoidea): A natural experiment in taphonomy. *Palaios*, **4**, 487–92.

Greenstein, B. J. (1990). Taphonomic biasing of subfossil echinoid populations adjacent to St. Croix, USVI. In D. K. Larue and G. Draper, eds., *12th Caribbean Geological Conference, August 7–11, 1989*. St. Croix, US Virgin Islands, pp. 290–300.

Greenstein, B. J. (1991). An integrated study of echinoid taphonomy: Predictions for the fossil record of four echinoid Families. *Palaios*, **6**, 519–40.

Greenstein, B. J. (1992). Taphonomic bias and the evolutionary history of the family Cidaridae (Echinodermata: Echinoidea). *Paleobiology*, **18**, 50–79.

Greenstein, B. J. (1993a). The effect of life habit on the preservation potential of echinoids. In B. N. White, ed., *Proceedings of the Sixth Symposium on the Geology of the Bahamas*. San Salvador, Bahamas: Bahamian Field Station, pp. 55–74.

Greenstein, B. J. (1993b). Is the fossil record of regular echinoids so poor? A comparison of living and subfossil assemblages. *Palaios*, **8**, 587–601.

Greenstein, B. J. (1995). The effects of life habit and test microstructure on the preservation potential of echinoids in Graham's Harbour, San Salvador Island, Bahamas. *Geological Society of America, Special Paper*, **300**, 177–88.

Grun, T. B. (2016). Echinoid test damage by a stingray predator. *Lethaia*, **49**, 285–86.

Grun, T. B. (2017). Recognizing traces of snail predation on the Caribbean sand dollar *Leodia sexiesperforata. Palaios*, **32**, 448–61.

Grun, T. B. and Nebelsick, J. H. (2016). Taphonomy of a clypeasteroid echinoid using a new quasimetric approach. *Acta Palaeontologica Polonica*, **61**, 689–99.

Grun, T. B. and Nebelsick, J. H. (2018a). Biomechanics of an echinoid's trabecular system. *PLoS ONE*, **13(9)**: e0204432.

Grun, T. B. and Nebelsick, J. H. (2018b). Technical biology of the clypeasteroid *Echinocyamus pusillus*: A review with outlook. *Contemporary Trends in Geoscience*, **7**, 247–54.

Grun. T. B. and Nebelsick, J. H. (2018c). Structural design analysis of the minute clypeasteroid echinoid *Echinocyamus pusillus. Royal Society Open Science*, **5**, 171323.

Grun, T. B., Koohi, L., Schwinn, T., et al (2016). The skeleton of the sand dollar as a biological role model for segmented shells in building construction: A research review. In J. Knippers, K. Nickel, and T. Speck, eds., *Biomimetic Research for Architecture and Building Construction: Biological Design and Integrative Structures*. Basle: Springer, 222–47.

Grun, T. B., Kroh, A., and Nebelsick, J. H. (2017). Comparative drilling predation on time-averaged phosphatized and non-phosphatized specimens of the minute clypeasteroid echinoid *Echinocyamus stellatus* from Miocene offshore sediments (Globigerina Limestone Fm., Malta. *Journal of Paleontology*, **91**, 633–42.

Grun, T. B., Mancosu, A., Belaústegui, Z., and Nebelsick, J. H. (2018). *Clypeaster* taphonomy: A paleontological tool to identify stable structures in natural shell systems. *Neues Jahrbuch für Geologie und Paläontologie, Abhandlungen*, **288**, 189–202.

Grun, T. B., Mihaljević, M., and Webb G. E. (2020). Comparative taphonomy of deep-sea and shallow-marine echinoids of the genus *Echinocyamus*. *Palaios*, **35**, 403–20.

Grun, T. B., Sievers, D., and Nebelsick, J. H. (2014). Drilling predation on the clypeasteroid echinoid *Echinocyamus pusillus* from the Mediterranean Sea (Giglio, Italy). *Historical Biology*, **26**, 745–57.

Guidetti, P. and Mori, M. (2005). Morpho-functional defenses of Mediterranean sea urchins, *Paracentrotus lividus* and *Arbacia lixula*, against fish predators: *Marine Biology*, **147**, 797–802.

Heatfield, B. M. (1971). Growth of the calcareous skeleton during regeneration of the spines of the sea urchin, *Strongylocentrotus purpuratus* (Stimpson): A light and scanning electron microscopic study. *Journal of Morphology*, **124**, 57–90.

Hendler, G. (1977). The differential effects of seasonal stress and predation on the stability of reef-flat echinoid populations. In D. L. Taylor, ed., *Proceedings of the 3rd International Coral Reef Symposium*, **1**. Miami, Florida: Rosenstiel School of Marine and Atmospheric Science, University of Miami, pp. 217–23.

Hess, H. (1972): Eine Echinodermen-Fauna aus dem mittleren Dogger des Aargauer Juras. *Schweizer Paläontologische Abhandlungen*, **92**, 1–87.

Hopkins, T. S., Thompson, L. E., Walker, J. M., and Davis, M. (2004). A study of epibiont distribution on the spines of the cidaroid sea urchin, *Eucidaris tribuloides* (Lamarck, 1816) from the shallow shelf of the eastern Gulf of Mexico. In T. Heinzeller and J. H. Nebelsick, eds., *Echinoderms München. Proceedings of the 11th International Echinoderm Meeting*. Rotterdam: Taylor & Francis, pp. 207–11.

Jagt, J. W. M., Dortangs, R., Simon, E., and van Knippenberg, P. (2007). First record of the ichnofossil *Podichnus centrifugalis* from the Maastrichtian of northeast Belgium. *Bulletin de l'Institut Royal des Sciences Naturelles de Belqique, Sciences de la Terre, Bulletin van het Koninklijk Belgisch Instituut vorr Natuurwetenschappen*, **77**, 95–105.

Jangoux, M. (1984). Diseases of echinoderms. *Helgoländer Meeresunte rsuchungen*, **37**, 207–16.

Johansson C. L., Bellwood D. R., Depczynski M., and Hoey, A. S. (2013). The distribution of the sea urchin *Echinometra mathaei* (de Blainville) and its predators on Ningaloo Reef, Western Australia: The implications for top-down control in an intact reef system. *Journal of Experimental Marine Biology and Ecology*, **442**, 39–46.

Johnson, A. S., Ellers, O., Lemire, J., Minor, M., and Leddy, H. A. (2002). Sutural loosening and skeletal flexibility during growth: Determination of drop-like shapes in sea urchins. *Proceedings of the Royal Society of London B*, **269**, 215–20.

Kidwell, S. M. and Baumiller, T. (1990). Experimental disintegration of regular echinoids: Roles of temperature, oxygen, and decay thresholds. *Paleobiology*, **16**, 247–71.

Kier, P. M. (1977). The poor fossil record of the regular echinoid. *Paleobiology*, **3**, 168–74.

Kier, P. M. (1981). A bored Cretaceous echinoid. *Journal of Paleontology*, **55**, 656–59.

Kowalewski, M. and Nebelsick, J. H. (2003). Predation on recent and fossil echinoids. In P. H. Kelley, M. Kowalewski, and T. A. Hansen, eds., *Predator-prey interactions in the fossil record*. Topics in Geobiology, 20. New York: Kluwer Academic/Plenum Publishers, pp. 279–302.

Kowalewski, M., Casebolt, S., Hua, Q., et al. (2018). One fossil record, multiple time resolutions: Disparate time-averaging of echinoids and mollusks on a Holocene carbonate platform. *Geology*, **46**, 51–54.

Krainer, K., Mostler, H. and Haditsch, J.G. (1994). Jurassische Beckenentwicklung in den Nördlichen Kalkalpen bei Lofer (Salzburg) unter besonderer Berücksichtigung der Manganerz-Genese. *Abhandlungen der geologischen Bundesanstalt*, **50**, 257–93.

Kroh, A. and Nebelsick, J. H. (2003). Echinoid assemblages as a tool for palaeoenvironmental reconstruction – an example from the early Miocene of Egypt. *Palaeogeography, Palaeoclimatology, Palaeoecology*, **201**, 157–77.

Kroh, A. and Nebelsick, J. H. (2010). Echinoderms and Oligo-Miocene carbonate system: Potential application in sedimentology and environmental reconstruction. *International Association of Sedimentologists, Special Publications*, **42**, 201–28.

Kudrewicz, R. (1992). The endemic echinoids *Micraster* (*Micraster*) *maleckii* Mączyńska, 1979, from the Santonian deposits of Korzkiew near Cracow (southern Poland); their ecology, taphonomy and evolutionary position. *Acta Geologica Polonica*, **42**, 124–34.

Kurz, R. C. (1995). Predator-prey interactions between Gray Triggerfish (*Balistes capriscus* Gmelin) and a guild of sand dollars around artificial reefs in the northeastern Gulf of Mexico. *Bulletin of Marine Science*, **56**, 150–60.

Kutscher, F. (1970). Die Echinodermen des Hunsrückschiefer-Meeres. *Abhandlungen des Hessischen Landesamtes für Bodenforschung*, **56**, 37–48.

Lawrence J. M. ed. (2020). (4th ed.). London: Academic Press, p. 718

Lawrence, J. M. and Vasquez, J. (1996). The effects of sublethal predation on the biology of echinoderms. *Oceanologica Acta*, **19**, 431–40.

Lewis, R. (1980). Taphonomy. In T. W. Broadhead and J. A. Waters, eds., *Echinoderms: Notes for a Short Course.* Studies in Geology, **3**, Knoxville: University of Tennessee Press, pp. 27–39.

Linse, K., Walker, L. J. and Barnes D. K. A. (2008). Biodiversity of echinoids and their epibionts around the Scotia Arc, Antarctica. *Antarctic Science*, **20**, 227–44.

Luidius, E. (1699). *Lithophylacii Britannicii ichnographia*. London.

McClanahan, T. R. (1988). Coexistence in a sea urchin guild and its implications to coral reef diversity and degradation. *Oecologia*, **77**, 210–18.

McClanahan, T. R. (1995). Fish predators and scavengers of the sea urchin *Echinometra mathaei* in Kenyan coral-reef marine parks. *Environmental Biology of Fishes*, **43**, 187–93.

McClanahan, T. R. (1998). Predation and the distribution and abundance of tropical sea urchin populations. *Journal of Experimental Marine Biology and Ecology*, **221**, 231–55.

McKinney, F. K. and Jackson, J. B. C. (1989). *Bryozoan Evolution*. Boston: Unwin-Hyman.

Märkel, K. and Röser, U. (1983). Calcite-resorption in the spine of the echinoid *Eucidaris tribuloides. Zoomorphology*, **103**, 43–58.

Mancosu, A. and Nebelsick, J. H. (2013). Multiple routes to mass accumulations of clypeasteroid echinoids: A comparative analysis of Miocene echinoid beds of Sardinia. *Palaeogeography, Palaeoclimatology, Palaeoecology*, **374**, 173–86.

Mancosu, A. and Nebelsick, J. H. (2015). The origin and paleoecology of clypeasteroid assemblages from different shelf settings of the Miocene of Sardinia, Italy. *Palaios*, **30**, 273–87.

Mancosu, A. and Nebelsick, J. H. (2016). Echinoid assemblages from the early Miocene of Funtanazza (Sardinia): A tool for reconstructing depositional environments along a shelf gradient. *Palaeogeography, Palaeoclimatology, Palaeoecology*, **454**, 139–60.

Mancosu, A. and Nebelsick, J. H. (2017a). Ecomorphological and taphonomic gradient of clypeasteroid-dominated echinoid assemblages along a mixed siliciclastic-carbonate shelf from the early Miocene of northern Sardinia, Italy. *Acta Palaeontologica Polonica*, **62**, 627–46.

Mancosu, A. and Nebelsick, J. H. (2017b). Palaeoecology and taphonomy of spatangoid-dominated echinoid assemblages: A case study from the early middle Miocene of Sardinia, Italy. *Palaeogeography, Palaeoclimatology, Palaeoecology*, **466**, 334–52.

Mancosu, A. and Nebelsick, J. H. (2019). Reconstructing the palaeoecology of echinoid dominated sublittoral environments: A case study from the Miocene of Sardinia. *Journal of Paleontology*, **93**, 764–84.

Mancosu, A. and Nebelsick, J. H. (2020). Tracking the preservation potential of regular sea urchins in recent and fossil shallow water, high energy environments. *Palaeontologia Electronica*. **23(2)**, a42.

Mancosu, A., Nebelsick, J. H., Kroh, A. and Pillola, G. L. (2015). The origin of echinoid shell beds in siliciclastic shelf environments: Three examples from the Miocene of Sardinia, Italy. *Lethaia*, **48**, 83–99.

Martinelli Filho, J. E., dos Santos, R. B., Ribeiro, C. C., (2014). Host selection, host-use pattern and competition in *Dissodactylus crinitichelis* and *Clypeasterophilus stebbingi* (Brachyura: Pinnotheridae). *Symbiosis*, **63**, 99–110.

Marcopoulos-Diacantoni, A. (1970). Some observations on the anomalies and irregularities of the test of echinoids, especially those from the Neogene of Greece (in Greek with a French summary). *Annales Géologiques des Pays Helléniques*, **22**, 256–62.

Marcopoulos-Diacantoni, A. (1984). Le genre *Clypeaster* dans le domaine Héllenique Durant le Néogène au point de vue Biostratigraphique – Paléoécologique – Taphonomique. *Annales Géologiques des Pays Helléniques*, **32**, 245–56.

Meadows, C. A., Fordyce, R. E. W., and Baumiller, T. K. (2015). Drill holes in the irregular echinoid, *Fibularia*, from the Oligocene of New Zealand. *Palaios*, **30**, 810–17.

Merrill R. J. and Hobson, E. S. (1970). Field Observations of *Dendraster excentricus*, a sand dollar of western North America. *American Midland Naturalist*, **83**, 595–624.

Mitrović-Petrović, J. (1972). Les apparitions des irrégularités et des anomalies sur le squelette des echinides du Miocene Moyen, comme la consequense du parasitisme et des lesions biotiques. (In Serbian with a French summary). *Geoloski anali balkanskoga poluostrva*, **31**, 135–45.

Mitrović-Petrović, J. (1982). Etudes taphonomiques du gisement contenant la faune des échinides (L´Èocene d´Istrie). In F. W. E. Rowe, ed., Papers from the Echinoderm Conference, The Australian Museum Sydney 1978. *Australian Museum Memoir*, **16**, 9–16.

Mitrović-Petrović, J. and Urošević-Dačić, D. (1962). Incrustings of bryozoan colonies on the shells of Middle Miocene echinoids. *Vesnik Zavoda za Geloška i Geofizička Istraživanja Series A*, **20**, 259–87.

Mizuno, Y. (1993). Echinoidea. In F. Ohe, I. Nonogaki, T. Tanaka, K. Hachiya, Y. Mizuno, T. Momoyama and T. Yamaoka, eds., *Fossils from the Miocene Morozaki Group*. Nagoya, Japan: Tokai Fossil Society,pp. 141–55.

Moffat, H. A. and Bottjer, D. J. (1999). Echinoid concentration beds: Two examples from the stratigraphic spectrum. *Palaeogeography, Palaeoclimatology, Palaeoecology*, **149**, 329–48.

Mortensen, T. (1934). Note on some fossil echinoids. *Geological Magazine*, **71**, 393–407.

Mortensen, T. (1937). Some echinoderm remains from the Jurassic of Württemberg. Kongelige Danske Videnskabernes Selskab, *Biologiske Meddelelser*, **13**, 1–28.

Mostler, H. (2009). Pedicellarien spät-norischer Echiniden aus der hallstätter Tiefschwellen-fazies, Nördliche Kalkalpen. *Geo.Alp*, **6**, 19–52.

Müller, A. H. (1957). *Lehrbuch der Paläozoologie. Band 1: Allgemeine Grundlagen*. Jena: VEB Gustav Fischer Verlag.

Nebelsick, J. H. (1992a). Echinoid distribution by fragment identification in the Northern Bay of Safaga, Red Sea, Egypt. *Palaios*, **7**, 316–28.

Nebelsick, J. H. (1992b). The Northern Bay of Safaga (Red Sea, Egypt): An actuopalaeontological approach. III, Distribution of echinoids: *Beiträge zur Paläontologie von Österreich*, **17**, 5–79.

Nebelsick, J. H. (1995a). The uses and limitations of actuopalaeontological investigations on echinoids. *Geobios,* Mémoire spéciaux, **18**, 329–336.

Nebelsick, J. H. (1995b). Comparative taphonomy of Clypeasteroids. *Eclogae Geologicae Helvetiae*, **88**, 685–93.

Nebelsick, J. H. (1995c). Actuopalaeontological investigations on echinoids: The potential for taphonomic interpretation. In R. H. Emson, A. B. Smith, and A. C. Campbell, eds., *Echinoderm Research*. Rotterdam: A. A. Balkema, pp. 209–14.

Nebelsick, J. H. (1996). Biodiversity of shallow-water Red Sea echinoids: implications for the fossil record. *Journal of the Marine Biological Association UK*, **76**, 185–94.

Nebelsick, J. H. (1999a). Taphonomic comparison between recent and fossil sand dollars. *Palaeogeography, Palaeoclimatology, Palaeoecology*, **149**, 349–58.

Nebelsick, J. H. (1999b). Taphonomic legacy of predation on echinoids. In M. D. Candia Carnevali and F. Bonasoro, eds., *Echinoderm Research 1998*. Rotterdam: A. A. Balkema, pp. 347–52.

Nebelsick, J. H. (1999c). Taphonomy of *Clypeaster* fragments: preservation and taphofacies. *Lethaia*, **32**, 241–52.

Nebelsick, J. H. (2004). Taphonomy of echinoderms: introduction and outlook. In T. Heinzeller and J. H. Nebelsick, eds., *Echinoderms München.*

Proceedings of the 11th International Echinoderm Meeting. Rotterdam: Taylor & Francis, pp.471–78.

Nebelsick, J. H. (2008). Taphonomy of the irregular echinoid *Clypeaster humilis* from the Red Sea: Implications for taxonomic resolution along taphonomic grades. In W. I. Ausich and G. D. Webster, eds., *Echinoderm Paleobiology*. Bloomington, IN: Indiana University Press, pp. 115–28.

Nebelsick, J. H. (2020). Clypeasteroids. In J. M. Lawrence, ed., *Biology and Ecology of Sea Urchins*, 4th ed. London: Academic Press, pp. 315–31.

Nebelsick, J. H. and Kampfer, S. (1994). Taphonomy of *Clypeaster humilis* and *Echinodiscus auritus* from the Red Sea. In B. David, A. Guille, J. P. Féral, and M. Roux, eds., *Echinoderms through Time*. Rotterdam: A. A. Balkema, pp. 803–08.

Nebelsick, J. H. and Kowalewski, M. (1999). Drilling predation on recent clypeasteroid echinoids from the Red Sea. *Palaios*, **14**, 127–44.

Nebelsick, J. H. and Kroh, A. (2002). The stormy path from life to death assemblages: The formation and preservation of mass accumulation of fossil sand dollars. *Palaios*, **17**, 378–93.

Nebelsick, J. H., Schmid, B., and Stachowitsch, M. (1997). The encrustation of fossil and recent sea-urchin tests: Ecological and taphonomical significance. *Lethaia*, **30**, 271–84.

Nebelsick, J. H., Dynowski, J. F., Grossmann, J. N., and Tötzke, C. (2015). Echinoderms: Hierarchically organized light weight skeletons. In C. Hamm, ed., *Evolution of Lightweight Structures: Analyses and Technical Applications*, Biologically-Inspired Systems, **6**. Basle: Springer Verlag, pp. 141–56.

Neumann, C. and Wisshak, M. (2006). A foraminiferal parasite on the sea urchin *Echinocorys*: Ichnological evidence from the Late Cretaceous (Lower Maastrichtian, northern Germany). *Ichnos*, **13**, 185–90.

Neumann, C. and Wisshak, M. (2009). Gastropod parasitism on Late Cretaceous to Early Paleocene holasteroid echinoids – evidence from *Oichnus halo* isp. n. *Palaeogeography, Palaeoclimatology, Palaeoecology*, **284**, 115–19.

Neumann, C., Wisshak, M., and Bromley, R. G. (2008). Boring a mobile domicile: An alternative to the conchicolous life habit. In M. Wisshak and L. Tapanila, eds., *Current Developments in Bioerosion*. Berlin-Heidelberg: Springer, pp. 307–28.

Perricone, V., Grun, T. B., Marmo, F., Langella, C., and Candia Carnevali, M. D. (2021). Constructional design of echinoid endoskeleton: Main structural components and their potential for biomimetic applications. *Bioinspiriration & Biomimetics*, **16**, 011001.

Petsios, E., Portell, R. W., Farrar, L., et al. (2021). An asynchronous Mesozoic marine revolution: The Cenozoic intensification of predation on echinoids. *Proceedings of the Royal Society B*, **288**, 20210400.

Peyer, K., Charbonnier, S., Allain, R., Läng, É., and Vacanta, R. (2014). A new look at the Late Jurassic Canjuers conservation Lagerstätte (Tithonian, Var, France). Nouveau regard sur le Lagerstätte de Canjuers, un site à conservation exceptionnelle du Jurassique supérieur (Tithonien, Var, France). *Comptes Rendus Palevol*, **13**, 403–20.

Philippe, M. (1983). Déformation d´une scutella (Echinoidea, Clypeasteroida) Miocène due à fixation d´une balane. Hypothèse paléoécologique. *Geobios*, **16**, 371–74.

Philippi, U. and Nachtigall, W. (1996). Functional morphology of regular echinoid tests (Echinodermata, Echinoida): A finite element study. *Zoomorphology*, **116**, 35–50.

Prouho, H. (1887). Recherches sur le *Dorocidaris papill*ata et quelques autres échinides de la Mediterranée. *Archives de zoologie expérimentale et générale*, **15**, 213–380.

Radwański, A. and Wysocka, A. (2001). Mass aggregation of Middle Miocene spine-coated echinoids *Echinocardium* and their integrated eco-taphonomy. *Acta Geologica Polonica*, **51**, 299–316.

Radwański, A. and Wysocka, A. (2004). A farewell to Świniary sequence of mass-aggregated, spine-coated echinoids *Psammechinus* and their associates (Middle Miocene; Holy Cross Mountains, Central Poland). *Acta Geologica Polonica*, **54**, 381–99.

Rahman, I. A., Belaústegui, Z., Zamora, S., et al. (2015). Miocene *Clypeaster* from Valencia (E Spain): Insights into the taphonomy and ichnology of bioeroded echinoids using X-ray micro-tomography. *Palaeogeography, Palaeoclimatology, Palaeoecology*, **438**, 168–79.

Roman, J. (1952). Quelques anomalies chez *Clypeaster melitensis* Michelin. *Bulletin de la Société Géologique de France*, **6**, 3–11.

Roman, J. (1953). Galles de myzostomides chez des clypéastres de Turquie. *Bulletin Muséum National d'Histoire Naturelle, Paris*, *2*, **25**, 287–313.

Roman, J. (1993). Taphonomie des échinodermes des calcaires lithographiques de Canjuers (Tithonien inférieur, Var, France). *Geobios,* Mémoire spéciaux, **16**, 147–55.

Roman, J. and Fabre J, (1986). Un rivage à échinoïdes reguliers de la base, du Crétacé à Canjuers (Aiguines, Var). *Actes 111eme Congrès national des Sociétés savants, Poitiers, Paris*, *Section sciences*, **1** (Science de Terre). 147–58.

Roman, J., Vadet A., and Boullier A. (1991) Echinoïdes et brachiopodes de la limite Jurassique-Crétacé à Canjuers (Var, France). *Revue Paléobiologie*, **10**, 21–7.

Roman, J., Atrops, F., Arnaud, M., et al. (1994). Le gisement tithonien inférieur descalcaires lithographiques de Canjuers (Var, France): État actuel des connaissances. The Early Tithonian lithographic limestones from Canjuers (Var, France): Present state of knowledge. *Geobios*, **27**, 127–35.

Romano, M. (2013). "The vain speculation disillusioned by the sense": The Italian painter Agostino Scilla (1629–1700) called "The Discoloured", and the correct interpretation of fossils as "lithified organisms" that once lived in the sea. *Historical Biology*, **26**, 631–51.

Rose, E. P. F. (1976). Some observations on the recent holectypoid echinoid *Echinoneus cyclostomus* and their palaeoecological significance. *Thalassia Jugoslavica*, **12**, 299–306.

Rose, E. P. F. and Cross N. F. (1993). The chalk sea urchin *Micraster*: Microevolution, adaptation and predation. *Geology Today*, **5**, 179–86.

Rosenkranz, D. (1971). Zur Sedimentologie und Ökologie von Echinodermen-Lagerstätten. *Neues Jahrbuch für Geologie und Paläontologie. Abhandlungen*, 138: 221–58.

Sala, E. and Zabala, M. (1996). Fish predation and the structure of the sea urchin *Paracentrotus lividus* populations in the NW Mediterranean. *Marine Ecology Progress Series*, **140**, 71–81,

Santos, A. G. and Mayoral, E. J. (2008). Colonization by barnacles on fossil *Clypeaster*: An exceptional example of larval settlement. *Lethaia*, **41**, 317–32.

Santos, A. G., Mayoral, E., Muñiz, F., Bajo, I., and Adriaensens, O. (2003). Bioerosión en erizos irregulares (Clypeasteroidea) del Mioceno superior en el sector suroccidental de la Cuenca del Guadalquivir (Provincia de Sevilla). *Revista Española de Paleontología*, **18**, 131–41.

Schäfer, W. (1962). *Aktuo-Paläontologie nach Studien in der Nordsee*. Frankfurt am Main: Verlag Waldemar Kramer.

Schäfer, W. (1972). *Ecology and Palaeoecology of Marine Environments*. Chicago: University of Chicago Press.

Scilla, A. (1670). *La vana speculatzione disingannata dal senso*. Naples: Andrea Colicchia.

Schneider, C. L. (2003). Hitchhiking on Pennsylvanian echinoids: Epibionts on *Archaeocidaris*. *Palaios*, **18**, 435–44.

Schneider, C. L. (2010). Epibionts on Late Carboniferous through Early Permian echinoid spines from Texas, USA. In L. G. Harris, S. A. Boetger, C. W. Walker, and M. P. Lesser, eds., *Echinoderms 2006. Proceedings of the*

12th International Echinoderm Conference, Durham, 7–11 August 2006. New Hampshire, USA. Boca Raton: CRC Press, pp.71–76.

Schneider, C. L., Sprinkle, J., and Ryder, D. (2005). Pennsylvanian (Late Carboniferous) echinoids from the Winchell Formation, north-central Texas, USA. *Journal of Paleontology*, **79**, 745–62.

Schwarz, A. (1930). Ein Seeigelstachel-Gestein. *Natur und Museum*, **12**, 502–06.

Seilacher, A. (1970). Begriff und Bedeutung der Fossil-Lagerstätten. *Neues Jahrbuch für Geologie und Paläontologie, Monatshefte*, **1970**, 34–39

Seilacher, A. (1979). Constructional morphology of sand dollars. *Palaeobiology*, **5**, 191–221.

Seilacher, A., Reif, W. E., and Westphal, F. (1985). Sedimentological, ecological and temporal patterns of fossil Lagerstätten. *Philosophical transactions of the Royal Society of London, series B: Biological sciences*, **311**, 5–24.

Sievers, D. and Nebelsick, J. H. (2018). Fish predation on a Mediterranean echinoid: Identification and preservation potential. *Palaios*, **33**, 47–54.

Sievers, D., Friedrich, J.-P., and Nebelsick, J. H. (2014). A feast for crows: Bird predation on irregular echinoids from Brittany, France. *Palaios*, **29**, 87–94.

Simon, A., Poulicek, M., Machiroux, R., and Thorez, J. (1990). Biodegradation anaérobique des structures squelettiques en milieu marin: 1 – Approche morphologique. *Cahiers de Biologie Marine*, **31**, 95–105.

Smith, A. B. (1984). *Echinoid Palaeobiology*. London: : :George Allen and Unwin Limited, p. 199.

Smith, A. B. (1990). Echinoid evolution from the Triassic to Lower Liassic. *Cahiers Université Catholique de Lyon, Série Scientifique*, **3**, 79–117.

Smith, A. B. (2005). Growth and form in echinoids: The evolutionary interplay of plate accretion and plate addition. In D. E. G. Briggs, ed., *Evolving Form and Function: Fossils and Development: Proceedings of a Symposium Honoring Adolf Seilacher for His Contributions to Paleontology in Celebration of His 80th Birthday*. New Haven: Peabody Museum of Natural History, Yale University, pp. 181–93.

Smith, A. B. and Rader, W. L. (2009). Echinoid diversity, preservation potential and sequence stratigraphical cycles in the Glen Rose Formation (early Albian, Early Cretaceous), Texas, USA. *Palaeobiodiversity and Palaeoenvironments*, **89**, 7–52.

Smith, A. B., Morris, N. J., Gale, A. S., and Rosen, B. R. (1995). Late Cretaceous (Maastrichtian) echinoid-mollusc-coral assemblages and palaeoenvironments from a Tethyan carbonate platform succession, northern Oman Mountains. *Palaeogeography, Palaeoclimatology, Palaeoecology*, **119**, 155–68.

Smith, D. S., del Castillo, J., Morales, M. and Luke, B. (1990). The attachment of collagenous ligament to stereom in primary spines of the sea-urchin *Eucidaris tribuloides*. *Tissue Cell*, **22**, 157–76.

Tasnádi-Kubaska, A. (1962). Paläopathologie, Pathologie der vorzeitliche Tiere. Jena: VEB Gustav Fischer Verlag.

Tavani, G. (1935). Sulle anomalie negli ambulacri di alcuni. Echini del Miocene della Cirenaica. *Atti Processi Verbali della Società Toscana di Scienze Naturali in Pisa*, **44**, 119–23.

Taylor, P. D. and Wilson, M. A. (2002). A New Terminology for Marine Organisms Inhabiting Hard Substrates. *Palaios*, **17**, 522–25.

Telford, M. (1985a). Domes, arches and urchins: The skeletal architecture of echinoids (Echinodermata). *Zoomorphology*, **105**, 114–24.

Telford, M. (1985b). Structural analysis of the test of *Echinocyamus pusillus* (O. F. Müller). In B. F. Keegan and B. D. S. O'Conner, eds., *Proceedings of the 5th International Echinoderm Conference, Ireland 1984*. Rotterdam: A. A. Balkema, pp. 353–60.

Thompson, J. R. and Ausich, W. I. (2016). Facies distribution and taphonomy of echinoids from the Fort Payne Formation (late Osagean, early Viséan, Mississippian) of Kentucky. *Journal of Paleontology*, **90**, 239–49.

Thompson, J. R. and Denayer, J. (2017). Revision of echinoids from the Tournaisian (Mississippian) of Belgium and the importance of disarticulated material in assessing palaeodiversity. *Geological Journal*, **52**: 529–538.

Thompson, J. R., Crittenden, J., Schneider, C. L., and Bottjer, D. J. (2015). Lower Pennsylvanian (Bashkirian) echinoids from the Marble Falls Formation, San Saba, Texas, USA. *Neues Jahrbuch für Geologie und Paläontologie, Abhandlungen*, **276**, 79–89.

Thuy, B., Gale, A.S., and Reich M. (2011). A new echinoderm Lagerstätte from the Pliensbachian (Early Jurassic) of the French Ardenne. *Swiss Journal of Palaeontology*, **130**, 173–85.

Tyler, C. I., Dexter, T. A., Portell, R. W., and Kowalewski, M. (2018). Predation-facilitated preservation of echinoids in a tropical marine environment. *Palaios*, **33**, 478–86.

Wilson, M. A., Borszcz, T., and Zapoń, M. (2015). Bitten spines reveal unique evidence for fish predation on Middle Jurassic echinoids. *Lethaia*, **48**, 4–9.

Wysocka, A., Radwański, A., and Górka, M. (2001). Mykolaiv Sands in Opole Minor and beyond: Sedimentary features and biotic content of Middle Miocene (Badenian) sand shoals of Western Ukraine. *Geological Quarterly*, **56**, 475–92.

Young, M. A. L. and Bellwood, D. R. (2011). Diel patterns in sea urchin activity and predation on sea urchins on the Great Barrier Reef. *Coral Reefs*, **30**, 729–36.

Zachos, L. G. (2008). Preservation of echinoid fossils, Paleocene and Eocene of Texas. *Transactions of the Gulf Coast Association of Geological Societies*, **58**, 919–32.

Zachos, L. G. (2009). A new computational growth model for sea urchin skeletons. *Journal of Theoretical Biology*, **259**, 646–57.

Zamora, A., Mayoral, E., Vintaned, J. A. G., Bajo, S., and Espílez, E. (2008). The infaunal echinoid *Micraster*: Taphonomic pathways indicated by sclerozoan trace and body fossils from the Upper Cretaceous of northern Spain. *Geobios*, **41**, 15–29.

Zatoń, M. ł., Villier, L., and Salamon, M. A. (2007). Signs of predation in the Middle Jurassic of south-central Poland: Evidence from echinoderm taphonomy. *Lethaia*, **40**, 139–51.

Zinsmeister, W. J. (1980). Observations in the predation of the clypeasteroid echinoid, *Monophoraster darwini* from the Upper Miocene Enterrios Formation, Patagonia, Argentina. *Journal of Paleontology*, **54**, 910–12.

Złotnik, M. and Ceranka, T. (2005). Patterns of drilling predation of cassid gastropods preying on echinoids from the middle Miocene of Poland. *Acta Palaeontologica Polonica*, **50**, 409–28.

Acknowledgments

We thank Bill Ausich, Jeff Thompson, and a further reviewer for very helpful comments for improving this manuscript. Photographs of echinoids (Fig. 2) were taken by Wolfgang Gerber (University of Tübingen).

Cambridge Elements

Elements of Paleontology

About the Series

The Elements of Paleontology series is a publishing collaboration between the Paleontological Society and Cambridge University Press. The series covers the full spectrum of topics in paleontology and paleobiology, and related topics in the Earth and life sciences of interest to students and researchers of paleontology.

The Paleontological Society is an international nonprofit organization devoted exclusively to the science of paleontology: invertebrate and vertebrate paleontology, micropaleontology, and paleobotany. The Society's mission is to advance the study of the fossil record through scientific research, education, and advocacy. Its vision is to be a leading global advocate for understanding life's history and evolution. The Society has several membership categories, including regular, amateur/avocational, student, and retired. Members, representing some 40 countries, include professional paleontologists, academicians, science editors, Earth science teachers, museum specialists, undergraduate and graduate students, postdoctoral scholars, and amateur/avocational paleontologists.

Cambridge Elements

Elements of Paleontology

Elements in the Series

Equity, Culture, and Place in Teaching Paleontology: Student-Centered Pedagogy for Broadening Participation
Christy C. Visaggi

Computational Fluid Dynamics and its Applications in Echinoderm Palaeobiology
Imran A. Rahman

Understanding the Tripartite Approach to Bayesian Divergence Time Estimation
Rachel C. M. Warnock and April M. Wright

The Stratigraphic Paleobiology of Nonmarine Systems
Steven Holland and Katharine M. Loughney

Echinoderm Morphological Disparity: Methods, Patterns, and Possibilities
Bradley Deline

Functional Micromorphology of the Echinoderm Skeleton
Przemyslaw Gorzelak

Disarticulation and Preservation of Fossil Echinoderms: Recognition of EcologicalTime Information in the Echinoderm Fossil Record
William I. Ausich

Crinoid Feeding Strategies: New Insights From Subsea Video And Time-Lapse
David Meyer, Margaret Veitch, Charles G. Messing and Angela Stevenson

Phylogenetic Comparative Methods: A User's Guide for Paleontologists
Laura C. Soul and David F. Wright

Expanded Sampling Across Ontogeny in Deltasuchus motherali *(Neosuchia, Crocodyliformes): Revealing Ecomorphological Niche Partitioning and Appalachian Endemism in Cenomanian Crocodyliforms*
Stephanie K. Drumheller, Thomas L. Adams, Hannah Maddox and Christopher R. Noto

Testing Character Evolution Models in Phylogenetic Paleobiology: A Case Study with Cambrian Echinoderms
April Wright, Peter J. Wagner and David F. Wright

The Taphonomy of Echinoids: Skeletal Morphologies, Environmental Factors, and Preservation Pathways
James H. Nebelsick and Andrea Mancosu

A full series listing is available at: www.cambridge.org/EPLY

Printed in the United States
by Baker & Taylor Publisher Services